RAND NATIONAL DEFENSE RESEARCH INSTITUTE

T0146359

Expanding Operating and Support Cost Analysis for Major Programs During the DoD Acquisition Process

Legal Requirements, Current Practices, and Recommendations

Michael Boito, Tim Conley, Joslyn Fleming, Alyssa Ramos, Katherine Anania

Prepared for the Office of the Secretary of Defense

Approved for public release; distribution unlimited

For more information on this publication, visit www.rand.org/t/RR2527

Library of Congress Cataloging-in-Publication Data is available for this publication.
ISBN: 978-1-9774-0089-5

Published by the RAND Corporation, Santa Monica, Calif.
© Copyright 2018 RAND Corporation
RAND® is a registered trademark.

Support RAND
Make a tax-deductible charitable contribution at
www.rand.org/giving/contribute

www.rand.org

Preface

The Weapon System Acquisition Reform Act of 2009 established the Office of Cost Assessment and Program Evaluation (CAPE) in the Office of the Secretary of Defense and mandated a broad set of cost analysis duties, including conducting independent cost estimates and cost assessments of major defense acquisition programs at acquisition milestones. Subsequent laws have mandated additional cost analysis duties for CAPE, especially pertaining to program operating and support (O&S) costs, and have expanded CAPE duties to focus more on product support activities and costs. O&S costs are those incurred after a system has been delivered to the field; product support includes O&S and activities and costs during acquisition that affect the reliability, maintainability, availability, and O&S cost of the system. CAPE asked the RAND Corporation to assess CAPE's cost-analysis activities in these areas and recommend ways that consideration of O&S cost issues could be improved during the acquisition process. RAND assessed the cost analysis requirements pertaining to O&S costs by reviewing relevant laws and U.S. Department of Defense (DoD) guidance; assessed the resources available to conduct the analyses by determining numbers of cost-estimating personnel, reviewing data typically available to inform cost analyses and cost-estimating processes and timelines; conducted interviews with government and industry subject-matter experts; and reviewed literature to develop recommendations to improve the cost analysis of weapon system product support.

The research was conducted from September 2016 to October 2017. Some familiarity with DoD and its processes for weapon system acquisition and sustainment on the part of the reader is assumed. The report should be of interest to those concerned with cost analysis and decisionmaking on weapon system acquisition and sustainment in DoD.

This research was sponsored by the Office of Cost Assessment and Program Evaluation within the Office of the Secretary of Defense and conducted within the Acquisition and Technology Policy Center of the RAND National Defense Research Institute, a federally funded research and development center sponsored by the Office of the Secretary of Defense, the Joint Staff, the Unified Combatant Commands, the Navy, the Marine Corps, the defense agencies, and the defense Intelligence Community.

For more information on the RAND Acquisition and Technology Policy Center, see www.rand.org/nsrd/ndri/centers/atp, or contact the director (contact information is provided on the webpage).

Contents

Figures

Figures

Tables

Summary

In the acquisition process, program managers within U.S. Department of Defense (DoD) components have primary responsibility for managing major defense acquisition programs (MDAPs), with the Office of the Secretary of Defense (OSD) providing oversight. During this process, an independent cost analysis is required at milestone reviews. The major milestones are currently Milestone A, prior to the Technology Maturation and Risk Reduction Phase; Milestone B, prior to the Engineering and Manufacturing Development Phase; and Milestone C, prior to the Production and Deployment Phase. Since 2009, the Office of Cost Assessment and Program Evaluation (CAPE), which was established within OSD as part of the Weapon System Acquisition Reform Act (WSARA) of 2009, has had responsibility for conducting or approving independent cost estimates (ICEs) and independent cost assessments (ICAs) for all MDAPs and major subprograms in advance of Milestone A, B, or C approval.

Cost is a key consideration in managing MDAPs. Prior to technology maturation and risk-reduction efforts that begin at Milestone A, cost estimates inform analyses of alternatives to meet a mission need. After Milestone A, cost estimates can inform cost goals used to weigh cost against other program attributes, such as schedule and performance. Cost estimates are used to ensure that adequate resources are budgeted and programmed to allow successful execution of the program. Cost estimates inform decisions about the force structure DoD can afford—how many units it can afford to procure and sustain.

In this report, we assess the extent to which CAPE has fulfilled the requirements for operating and support (O&S) cost analyses of MDAPs mandated in WSARA and subsequent laws through the FY 2017 National Defense Authorization Act (NDAA; Public Law [Pub. L.] 114-328, 2016). We address the following questions:

- What are the legal requirements for CAPE and the military departments regarding O&S ICEs and ICAs?
- What is CAPE's O&S cost-estimating workload due to these requirements?
- What resources (personnel, data, time, etc.) are available to perform these duties?
- What duties required of CAPE regarding O&S costs are not accomplished due to resource constraints?
- How can CAPE improve its ability to meet the requirements?

In conducting this assessment, RAND reviewed relevant laws and DoD guidance, committee reports from the House and Senate Armed Services Committees, and testimony of witnesses before these committees. We obtained counts of CAPE cost estimates of MDAPs over several years and interviewed cost analysts from CAPE and the independent cost-estimating organizations in each military service to determine cost analysis workloads and methodologies. We obtained counts of cost-estimating personnel and reviewed data sources typically available to inform cost analyses, including service cost databases and documentation provided to cost analysts. We interviewed government and industry subject-matter experts (SMEs) who participate in the DoD acquisition process and provide inputs to cost analysts. We analyzed O&S cost estimates of MDAPs in Selected Acquisition Reports.

Findings

To understand DoD's experience and the concerns of stakeholders, we examined the estimated O&S cost per unit of selected MDAPs over the last 20 years. We selected the programs with the largest life-cycle O&S costs based on their Selected Acquisition Report estimates. These programs are not representative of all MDAPs or of all DoD programs.

Operating and Support Costs Are Often Underestimated, Especially in Early Phases

We found that constant-dollar estimates of O&S cost per unit for a few of these MDAPs doubled or tripled from the initial estimates. For these MDAPs, the technical inputs for reliability and maintainability (R&M) provided to cost estimators early in development often turned out to be overstated compared to actual results when the systems were fielded. The estimated life-cycle costs for these few programs totaled several hundreds of billions of FY 2017 dollars. We present the details of this analysis in Appendix A.

Many Systems Are Not Meeting Reliability and Availability Goals

We also found that fewer systems are meeting their reliability and availability goals when tested. Multiple DoD Directors of Operational Test and Evaluation (DOT&Es) have observed this trend and written about it in their annual reports. In FYs 2015 and 2016, for example, 36 percent and 42 percent of tested programs, respectively, met their reliability requirements (DOT&E, 2016a, DOT&E, 2016b). The directors assess operational suitability during testing, which includes an assessment of the availability, reliability, and maintainability of a system in its intended operational environment. They have observed a declining trend in the reliability and suitability of recently tested systems compared with those tested decades ago.

Legal Requirements for Estimates or Assessments of O&S Costs and Activities

Legal and regulatory requirements related to acquisition reform indicate an increasing emphasis on O&S costs and related logistics outcomes, such as R&M and availability. U.S. Code Title 10 §2334 requires CAPE to conduct or approve ICEs and ICAs for all MDAPs and major subprograms before Milestone A, B, or C approval. In addition, the law requires CAPE to review all cost estimates for and cost analyses of MDAPs and major subprograms.

Laws subsequent to WSARA's establishment of CAPE in 2009 added requirements at MDAP reviews. The new requirements include assessing the adequacy of funding for sustainment planning, conducting sensitivity analyses of key cost drivers affecting life-cycle costs, setting cost goals for procurement and O&S, and evaluating alternative courses of action that might reduce cost and risk. The FY 2017 NDAA (Pub. L. 114-328, 2016) added a requirement for the secretary of each military department to conduct periodic sustainment reviews of MDAPs, which are to include an independent estimate of the remainder of the program's life-cycle cost.

The additional duties for CAPE extend beyond estimation of *O&S* costs, which are those costs incurred after deployment of a system. The additional duties encompass consideration of *product-support* costs, which include O&S costs but also include costs for activities during development and procurement accomplished to ensure the fielded system is available, reliable, and affordable. These activities include system engineering and design and providing for maintenance, supply support, training, and other functions. The broadening of responsibilities from O&S to product-support activities and cost analyses is an expansion of traditional CAPE cost-analysis duties.

Our examination of the laws, House Armed Services Committee and Senate Armed Service Committee reports, testimony of experts before the committees, and statements from the leaders of these committees clearly indicate congressional concern about DoD system O&S cost and logistics outcomes and the intention to improve the outcomes through legislation.

CAPE's Operating and Support Cost Workload Exceeds Resources

From FY 2010 through FY 2016, CAPE averaged 16 cost-estimating events per year that required O&S cost estimates or assessments. Most of these were MDAP milestone reviews. These events did not include the new requirement for sustainment reviews, and the scope of the estimates did not include the tasks added in the FY 2016 (Pub. L. 114–92, 2015) and FY 2017 NDAAs.

CAPE has four dedicated O&S cost analysts. Independent analysts typically spend four to six months on each cost estimate. A large majority of an analyst's time is spent collecting information from SMEs on the system being analyzed, especially information on differences between the system and its antecedent that will drive O&S costs.

Available Information Sources Are Insufficient to Assess Product Support and O&S Costs

CAPE analysts draw on information in the cost-assessment requirements description (CARD) and, to a lesser extent, the life-cycle sustainment plan to inform their estimates. We reviewed these documents for several MDAPs and found they contained some useful information, but we appreciated why analysts needed to supplement this information as they do. In addition, we found that some CARDs for MDAPs that are now fielded and for which actual data are available had significantly overstated R&M metrics compared with actual results achieved on fielding.

The service Visibility and Management of Operating and Support Cost (VAMOSC) systems, which are the official service databases for system O&S costs, are of paramount importance in estimating O&S costs. Analysts typically estimate by analogy with an antecedent system, adjusting for significant differences, and the VAMOSC systems provide cost and programmatic data on these antecedent systems. We found that the Navy and Air Force VAMOSC systems generally reported the O&S costs defined in DoD's O&S cost element structure and provided an adequate representation of antecedent system costs. The Army VAMOSC system is less comprehensive, and the Army is taking steps to expand the capture and reporting of the cost of its systems. We found that a more complete reporting of the yearly quantities, costs, and scope of depot maintenance events would better allow cost analysts to assess and forecast this sizable element of O&S costs.

Data on Many Contractor Costs Are Lacking

For insight into contractor costs, cost analysts use cost reports submitted by contractors called *cost and software data reports*. Two kinds of contractor costs are of interest to O&S cost analysts. One cost of interest is planning for logistics support, which is funded and conducted during development and production. Elements include peculiar support equipment, peculiar training equipment, publications and technical data, and initial spares. We examined contractor cost reports for these efforts for several current MDAPs and found the reports offer little to no programmatic information that would make the data useful to cost analysts.

Another potential source for this information on legacy systems is the DoD organizations that provide the logistics support for the systems. We found no central repository for such historical data, and the cost analysts we interviewed reported finding it difficult or impossible to obtain the data. Because the logistics support efforts during acquisition affect O&S costs and logistics performance, the lack of insight into these efforts is a serious shortcoming.

A second contractor cost of interest to analysts is that for systems for which contractors provided logistics support. Again, the cost reports are almost entirely devoid of programmatic information useful to cost analysts, such as the number of maintenance actions or repair activities conducted. This information is critical because systems tend to require varying amounts of maintenance as they age. Without programmatic information, the analyst cannot determine trends related to age and usage.

Access to Subject-Matter Expertise Is Needed in Many Areas Related to O&S

Access to independent, substantive expertise in such areas as contracting, logistics, manpower, and reliability is another critical resource for cost analysts. We found that CAPE analysts often have difficulty obtaining substantive input from peers in OSD.

CAPE O&S Cost Duties Are Not Fully Accomplished Due to Resource Constraints

We found that CAPE's four O&S cost analysts cannot perform all the cost activities mandated in law or do them with the requisite analytical rigor. The analysts accomplish roughly one-half of the workload MDAP reviews generate. In addition, a reasonable interpretation of the requirements added in the FY 2016 and FY 2017 NDAAs that address planning for sustainment, establishing cost goals, and life-cycle estimates during sustainment reviews could double the previous historical workload.

Recommendations

Recommendations for Meeting CAPE's Statutory Responsibilities for O&S Cost Analysis

Our first set of recommendations is intended to fulfill a minimal interpretation of laws regarding CAPE O&S cost activities, assuming historical levels of effort and products.

Augment CAPE Staff

The number of additional CAPE O&S analysts needed to fulfill all statutory requirements depends on the rigor with which the tasks are done. Assuming a level of rigor consistent with historical CAPE ICEs, we estimate that CAPE needs between ten and 16 O&S analysts. We recommend hiring new staff at lower grade levels and developing their skills and experience through teaming with senior staff, which would facilitate longer tenure and less turnover. We recommend augmenting CAPE staff accordingly. In recognition of the pressure to reduce headquarters staff, billets could be transferred from former OSD Acquisition, Technology, & Logistics organizations.

Continue Management Support of Existing Efforts to Address Data Gaps and Make Additional Data Available to Support O&S Cost Analyses

CAPE and the services are already addressing weaknesses in such data sources as CARDs, life-cycle sustainment plans, cost and software data reports, and VAMOSC systems. In addition, the military components collect much information that would be useful for CAPE cost analysis but do not share the information with CAPE. This includes cost and programmatic information on depot maintenance, requirements for spare parts to achieve readiness objectives, and the serviceable inventory levels of the parts. The provisions in the FY 2018 NDAA (Pub. L. 115-91, 2017) for OSD to establish a common enterprise of business systems that extract data from relevant systems in DoD should make it easier for all participants in the acquisition process, including CAPE, to obtain information needed for decisionmaking and oversight. We recommend continued management support of these efforts.

We recommend making funding of VAMOSC systems an OSD responsibility to ensure that the services do not reduce budgeted levels during execution but having the services continue to manage the systems.

These recommendations to augment CAPE staff and ensure access to needed data are consistent with 10 U.S. Code 2334, the section of law that describes CAPE's duties and requires it to have enough staff and data access to perform those duties.

Recommendations for Improving O&S Outcomes in the Department of Defense

DoD efforts and legislation to date have improved O&S cost-estimating practices from those we found in our examination of estimates generated before CAPE was established. Our first set of recommendations is consistent with the preferences of DoD leadership and expressed in legislation for a minimal OSD oversight role. However, this approach is not as likely to fulfill the intent of recent legislation to improve cost and logistics outcomes as our second set of recommendations, which endorses more-continuous CAPE involvement with selected MDAPs.

Thus, our second set of recommendations is designed to fulfill the intent of legislation through the FY 2017 NDAA to improve O&S cost and logistics outcomes. Among the intents of the legislation is to ensure that, during the acquisition phase, DoD sets goals for O&S costs of MDAPs, plans for favorable cost and logistics outcomes, and ensures adequate funding for logistics support. These duties require a level of insight regarding the technical requirements for logistics planning and associated costs that a cost analyst alone cannot be expected to possess. Substantive expertise in system engineering and logistics planning, in addition to cost analysis, is needed.

In a program management organization, the SMEs in these disciplines would proceed through well-understood logistics engineering and planning steps to identify logistics support requirements and costs. Chapter Seven of this report cites evidence that DoD lost much of this expertise during drawdowns in the acquisition workforce and continues to experience shortfalls in expertise and that the cost and logistics outcomes have suffered as a result. We recommend a more robust OSD role for selected MDAPs that could fulfill legislative intent in this area. We realize the intent of Congress and DoD leadership is to reduce OSD's role in acquisition and recommend the following approach on a selected small portfolio of MDAPs.

Strengthen the OSD Role to Encourage Improved O&S Cost and Logistics Outcomes

To ensure the combination of expertise in cost, logistics, and systems engineering needed to identify, oversee, and ensure adequate resources for logistics planning during acquisition and estimate product-support costs, we recommend augmenting CAPE's staff with experts to provide inputs to its cost analysts. CAPE is uniquely positioned in DoD and empowered by law to provide independent and objective assessments and obtain the information it needs to do so. CAPE could hire the expertise permanently or contract for it from other independent organizations.

A less desirable option, which we offer in recognition of aversion to increasing headquarters staffs, is to establish an OSD process similar to Naval Air Systems Command's Estimating Technical Assurance Board process. The purpose of the process is to ensure that credible technical inputs are provided for cost estimates. The process requires general officer or senior executive service–level leaders in technical competencies to validate the technical inputs used in cost estimates at major reviews. In the process we envision, OSD offices with expertise in system engineering and logistics management and any other required substantive discipline would be required to provide technical inputs to CAPE analysts.

We concluded that such a role is needed for several reasons, including characteristics of the DoD acquisition process that inhibit system engineering for better O&S cost and logistics outcomes. These characteristics include relaxed requirements regarding system engineering for R&M and logistics support for DoD systems as a result of acquisition reform efforts, a diminished system engineering capability within the federal government, cultural barriers in DoD that inhibit cooperation and information sharing, and inadequate incentives in the acquisition system for improved O&S cost and logistics outcomes.

Make OSD Subject-Matter Experts Continuously Available to the Program

In addition to the recommendation to establish a process for providing technical inputs to CAPE estimates, we recommend that these OSD SMEs be continuously available to the program office to interact and lend expertise informally. Continuous, rather than episodic, involvement would have at least four benefits. First, it would allow sharing of OSD expertise that draws across DoD components and commodity types. Second, it would allow more-effective oversight. CAPE O&S analysts now have episodic involvement with MDAPs for a few months prior to a milestone review, which does not allow the analysts to monitor progress on efforts during development that affect O&S outcomes. Third, continuous informal interaction would lessen the need for often time-consuming formal documents and reviews, which are widely criticized for diverting management attention. Fourth, a formal requirement to combine substantive expertise would allow CAPE to effectively examine risk drivers, explore alternatives to the program of record, and generally test underlying or framing assumptions, which is the intent of recent legislation and was the intent when the capability for independent analysis was created in OSD in the early 1960s.

Acknowledgments

We are grateful to many people who helped us conduct this research. Thomas Henry of the Office of Cost Assessment and Program Evaluation (CAPE) sponsored this project. Donald Dawson of CAPE served as the action officer for the project and connected us with sources of information in the U.S. Department of Defense. Stephanie Patton of CAPE provided insights into recent statutory requirements for CAPE cost analysis. Larry Klapper and Molly Mertz of the Office of the Assistant Secretary of Defense for Logistics and Materiel Readiness provided documents and data associated with major defense acquisition programs and facilitated interviews with subject-matter experts in logistics, and we thank Terence Emmert of that office for making his staff available to us. Ranae Woods of the Air Force Cost Analysis Agency, Stephen Loftus of the Office of the Deputy Assistant Secretary of the Army for Cost Estimating, and Wendy Kunc of the Naval Center for Cost Analysis allowed us to interview their staffs and provided information on the staffing and cost-estimating activities of their organizations. John Wallace, Brian Phillips, and Angelena Bohman of the RAND Corporation conducted research and literature review. Cynthia Cook, Chris Mouton, and Brian Dougherty of RAND offered guidance, reviewed a draft of the report, and provided suggestions that improved our research. Erin-Elizabeth Johnson and Kristin Leuschner of RAND edited earlier drafts of the report, and Phyllis Gilmore performed the final edit. We thank unnamed subject-matter experts from industry for sharing their perspectives on product support of defense programs.

The content of the report is the sole responsibility of the authors.

CHAPTER ONE

The U.S. Department of Defense Cost Analysis Ecosystem and the Office of Cost Assessment and Program Evaluation

Cost is a key consideration in managing major defense acquisition programs (MDAPs) in the U.S. Department of Defense (DoD). In the acquisition process, program offices within DoD components have primary responsibility for managing programs, with oversight provided by the Office of the Secretary of Defense (OSD). Each program undergoes a series of milestone reviews, during which the milestone decision authority (the executive with overall responsibility for the program, usually either the defense acquisition executive or service acquisition executive) reviews the progress of a program and its suitability to proceed to the next phase of the acquisition life cycle. The major milestones are currently Milestone A, prior to the Technology Maturation and Risk Reduction Phase; Milestone B, prior to the Engineering and Manufacturing Development (EMD) Phase; and Milestone C, prior to the Production and Deployment Phase. The Director of Cost Assessment and Program Evaluation (CAPE), which lies within OSD and was established as part of the Weapon System Acquisition Reform Act of 2009 (WSARA), has responsibility for conducting or approving independent cost estimates (ICEs) and independent cost assessments (ICAs) for all MDAPs and major subprograms in advance of Milestone A, B, or C approval.[1] The Director of CAPE reports directly to the Secretary of Defense and is the secretary's principal adviser for independent cost assessment, program evaluation, and analysis.[2]

Figure 1.1 shows the generic phases of the defense acquisition process and the associated decisions that precede each phase. The law in 10 U.S.C. 2334 requires CAPE to either conduct or approve ICEs in support of Milestones A, B, and C for MDAPs.[3] The legal requirements for CAPE to conduct or review ICEs of MDAPs at these mile-

---

[1] An ICE or ICA is conducted by an organization not affiliated with the organization managing the program that is the subject of the estimate or assessment. Each military department has a cost-estimating organization independent of the program offices that manage programs.

[2] The Director of CAPE is appointed by the President, and the position and its responsibilities are defined in 10 U.S. Code Title 10 (10 U.S.C.) 139a.

[3] As defined in 10 U.S.C. 2430, an MDAP is a program that is estimated to cost more than $300 million for development or more than $1.8 billion for procurement (in fiscal year [FY] 1990 constant dollars).

Figure 1.1
Generic Phases in Defense Acquisition Programs and Major Decisions

Generic defense program phase	Materiel solution analysis	Technology maturation and risk reduction	EMD	Production and deployment	O&S

Decision	Materiel development	Milestone A Risk reduction	Milestone B Development contract award	Milestone C Initial production or fielding	N/A

SOURCE: DoD Instruction (DoDI) 5000.02, 2015, Figures 1 and 2.
RAND RR2527-1.1

stone decisions drive most of the cost-analysis activities in CAPE. We discuss the legal requirements in more detail in Chapter Two, and CAPE's cost-analysis activities and workload in Chapter Three.

In this report, we assess the extent to which CAPE has fulfilled the requirements for operating and support (O&S) cost analyses of MDAPs mandated in WSARA and subsequent laws through the FY 2017 National Defense Authorization Act (NDAA).[4] We address the following questions:

- What are the legal requirements for CAPE and the military departments regarding O&S cost estimating and cost assessment?
- What is CAPE's O&S cost-estimating workload due to these requirements?
- What resources (personnel, data, time, etc.) are available to perform these duties?
- What duties required of CAPE regarding O&S costs are not accomplished due to resource constraints?
- How can CAPE improve its ability to meet the requirements?

Approach and Data Sources

To address these research questions, we reviewed relevant laws and DoD guidance. We read committee reports from the House and Senate Armed Services Committees and testimony of witnesses before the committees to better understand the intent of

---

[4]   CAPE's role in O&S cost analyses is only part of its cost-analysis role and one of its functions. CAPE has cost analysis duties for the entire life cycle of MDAP costs. CAPE also supports resource planning in DoD's planning, programming, budgeting, and execution process; conducts analyses in support of the planning phase of the process; prepares programmatic guidance for the Future Years Defense Program; and manages the program-review phase of the planning, programming, budgeting, and execution process. CAPE also provides analysis and advice regarding requirements being considered by the Joint Requirements Oversight Council.

the legislation. We obtained counts of CAPE cost estimates of MDAPs from 2010 to 2016 and conducted structured interviews with cost analysts from CAPE and the independent cost-estimating organizations in each military service to understand the tasks, methodologies, and resources associated with the duties of independent cost analysts. We obtained counts of cost-estimating personnel and reviewed data sources typically available to inform cost analyses, including service cost databases and documentation provided to cost analysts. We analyzed O&S estimates of MDAPs in Selected Acquisition Reports (SARs) to assess changes in the estimates over time and, more important, reasons for changes greater than average. We drew on our interviews with government cost and logistics subject-matter experts (SMEs) and our review of law, regulations, and literature to develop recommendations to improve the cost analysis of MDAP O&S costs.

The remainder of this introduction discusses the reasons for increased concern about O&S costs for MDAPs and then provides a synopsis of OSD O&S cost analysis policies.

The Focus on O&S and CAPE O&S Cost Activities

DoD faces a difficult challenge in fielding systems that are reliable, maintainable, available, and affordable. Its systems must keep pace with an increasingly capable threat. As a result, DoD weapon systems have become more capable and more complex. The added capability and complexity of new systems come with increased cost. O&S costs, in particular, have been a growing source of concern. The cost to sustain existing weapon systems, such as aircraft, has outpaced the rate of inflation (Boito et al., 2016), and new weapon systems tend to cost more to sustain than the systems they replace (Kneece et al., 2014). Furthermore, many programs are not achieving their reliability and maintainability (R&M) targets, which contributes to increased life-cycle costs. These realities make it challenging for DoD to buy and sustain the forces it needs within its budget and highlight the need for accurate information about prospective acquisition program costs.

Congressional concern about the trends in O&S costs is apparent in WSARA and analysis conducted shortly thereafter by the Government Accountability Office (GAO) and the Congressional Research Service.[5] In the 2009 legislation, Congress directed GAO to submit to the congressional defense committees a report on growth in O&S costs for major weapon systems. GAO was directed to, among other tasks, analyze the rate of growth for O&S costs for major weapon systems, assess causes of the growth, and assess measures DoD has taken to reduce the costs.

---

[5]   Before 2004, GAO was known as the General Accounting Office.

The legislation also directed CAPE to review its capabilities to track and assess O&S costs on major programs and assess the feasibility and advisability of establishing baselines for O&S costs for major programs (Public Law [Pub. L.] 111-23, 2009). Current law requires DoD to set goals for procurement unit costs and sustainment costs early in system development, establish baselines for acquisition unit costs, and report breaches of these cost baselines to Congress.

GAO's 2010 report in response to the direction in WSARA stated that "DOD lacks key information needed to effectively manage and reduce O&S costs for most of the weapon systems GAO reviewed—including life-cycle O&S cost estimates and complete historical data on actual O&S costs" (GAO, 2010). GAO also found that the military services did not regularly update the O&S cost estimates after production had been completed for six of the seven systems analyzed by GAO.

Growing Concern over Other O&S Outcomes

During the same period that DoD nonpay operating costs were increasing, there was mounting evidence of other undesirable outcomes in the O&S phase of DoD systems, such as declining trends in reliability, maintainability, and availability. *Availability* is the percentage of time a system can perform its wartime mission. Several conditions can cause a weapon system to be unavailable, including lack of spare parts and time spent in maintenance at the unit level or in a depot. Test results indicated that the availability of new weapon systems in general was lower than their predecessors at the same point in the life cycle and that the availability of several high-profile weapon systems was far below goals established for them early in development.

Worrisome trends in weapon system reliability, maintainability, and availability have been identified consistently by different DoD Directors of Operational Test and Evaluation (DOT&Es) for more than a decade. DOT&E assesses the operational effectiveness and suitability of weapon systems. Suitability includes an assessment of the availability, reliability, and maintainability of a system in its intended operation environment. Director Thomas P. Christie wrote in the FY 2004 DOT&E report:

> In the history of DOT&E reports to Congress since 1983, about 30 percent of systems (36 of 126) were less than suitable. Recent years have witnessed an increase in the number of systems found unsuitable in operational testing and evaluation. Suitability problems add significantly to the logistics burden and life cycle costs of programs. The Defense Science Board [DSB] in 2000 pointed out that 80 percent of defense systems brought to operational test fail to achieve even half of their reliability requirement. (DOT&E, 2004, pp. i–ii)

Director Charles E. McQueary wrote in the FY 2007 DOT&E report:

DOT&E has sent a total of 144 system reports to Congress since 1983 and we assessed 103 of the systems as suitable (72 percent). This past year's result of 50 percent reveals a continued downward trend. (DOT&E, 2007, p. i).

Director J. Michael Gilmore wrote in the FY 2013 DOT&E report:

From FY97 to FY13, 56 percent (75 of 135) of the systems that conducted an operational test met or exceeded their reliability threshold requirements as compared to nearly 64 percent between FY85 and FY96. (DOT&E, 2014, p. vi)

Among the programs with suitability issues identified in DOT&E reports since FY 2005 are the F-22 fighter aircraft, CV-22 and MV-22 tilt-rotor aircraft, Global Hawk remotely piloted aircraft, littoral combat ship (LCS), and F-35 strike fighter aircraft programs.

Results of our analysis of the O&S cost estimates for these and other programs, reported in Appendix A, show that, along with suitability problems identified in DOT&E reports, the programs also had much higher than average growth in their estimated unit O&S cost compared with estimates made early in the development phases of these programs.[6] The estimates of unit O&S costs of some of these large programs doubled or tripled in real terms. This is not a coincidence. As the DOT&Es quoted earlier highlight in their annual reports, R&M are linked to O&S costs, and all these outcomes are strongly influenced early in development. SMEs from organizations including DOT&E, the National Research Council (NRC), and DSB have argued that these unfavorable outcomes are due, at least in part, to changes in DoD policies and workforce levels. We discuss these issues more in Chapter Seven.

Congress's concern about O&S cost and suitability outcomes and its intention to address these issues are clear in the legislation summarized in Chapter Two. WSARA unified OSD's cost-estimating and programming function into CAPE and specified its initial duties. Congress further specified CAPE and departmental responsibilities for O&S cost activities and management of other sustainment outcomes in subsequent legislation.

Although the O&S phase, which begins with deployment of the system to the field, is not preceded by a milestone decision separate from Milestone C, recent legislation, especially in the FY 2012, FY 2016, and FY 2017 NDAAs, has added requirements for analysis of product support and O&S costs in support of MDAP milestone decisions. These include assessing the adequacy of funding for sustainment planning, conducting sensitivity analyses of key cost drivers affecting life-cycle costs, setting cost goals for procurement and O&S, and evaluating alternative courses of action that may reduce cost and risk.

---

[6] The programs are at different points in their development and fielding, and there is the potential for their actual and estimated O&S costs to change from current estimates.

Synopsis of OSD O&S Cost Analysis Experience

This section briefly explores the history of OSD's O&S cost analysis experience and discusses the establishment of CAPE and the evolution of cost analysis policies for MDAPs within DoD, particularly regarding O&S costs.

Establishment of CAPE

CAPE traces its roots to management techniques introduced to OSD by then–Secretary of Defense Robert McNamara and his staff in the 1960s. While serving as secretary from 1961 to 1968, McNamara instituted the Planning, Programming, and Budgeting System to centralize these processes, which aimed to provide a more thorough, analytical, and systematic way of making decisions about force structure, weapon systems, and costs.[7] To help implement the new management system, an Office of Systems Analysis was established within the Comptroller Office (Fisher, 1970).

With the increased responsibilities given to a new staff of systems analysts in OSD, the military services found that their programs and budgets were coming under more scrutiny. Service leaders tended to resent "what they considered intrusion on their traditional prerogatives" (Trask and Goldberg, 1997, p. 34), and, by the late 1960s, military leaders were publicly criticizing OSD's role in analyzing and making decisions based on the cost-effectiveness of proposed systems (Hough, 1989).

After Melvin Laird became Secretary of Defense in 1969, OSD instituted changes to the acquisition process, returning to the military services the responsibility for identifying needs for weapon systems and for defining, developing, and producing the systems. The Defense Systems Acquisition Review Council was established within OSD to advise the Deputy Secretary of Defense on the status and readiness of MDAPs to proceed through each phase of the acquisition life cycle. Policy guidance stated that DoD components were responsible for identifying needs for defense systems and for acquiring them, and management oversight and reporting requirements should be kept to a minimum (DoD Directive [DoDD] 5000.1, 1971). The directive instructed components to request OSD approval to proceed through the acquisition milestones, subject to meeting specified criteria. The same basic process is used today, although reporting requirements have grown since the early 1970s.

In support of the Defense Systems Acquisition Review Council, the Cost Analysis Improvement Group (CAIG) was established in 1972 to provide ICEs and to establish uniform DoD cost-estimating standards for use throughout DoD (Senate Committee on Armed Services, 1985). The group performed this role until it was reorganized into CAPE in 2009.

---

[7] For insights into changes in organization and management in DoD and the problems these changes tried to address, see Trask and Goldberg (1997) and Fox et al. (2011).

DoD Policy Regarding Cost in Acquisition of Major Defense Systems
Notable from a cost perspective is that the acquisition policy guidance issued in 1971 (as DoDD 5000.1, 1971) stipulated that life-cycle costs would be considered at milestone decisions to enter development and production or deployment, that cost goals that include the cost of acquisition and sustainment should be established, and that discrete costs (such as unit production cost and operating and support cost) should be translated into design-to-cost (DTC) requirements. Also notable from a sustainment cost perspective was the guidance to consider logistics support a design parameter (DoDD 5000.1, 1971).

A separate directive on the subject, DoDD 5000.28, was issued to DoD components in 1975, clarifying that the intent of the policy was to establish cost as equal in importance to performance and schedule and to establish cost parameters as management goals for program managers and contractors to balance cost, performance, and schedule. O&S costs were to be considered. The directive acknowledged the inability to estimate O&S costs as rigorously as procurement costs but emphasized that controlling future O&S costs should be a management goal and that goals for parameters that affect O&S cost (e.g., the number of maintenance personnel, R&M metrics) should be established. The directive also stipulated that efforts to improve available data on O&S costs would continue.

As of fall 2017, DoD acquisition policy stipulates that trade-offs among cost, schedule, and performance be made throughout the program life cycle and assigns this responsibility to the secretaries of the military departments, service chiefs, and program managers (DoDI 5000.02, 2017). DoD acquisition policies have continued to mandate consideration of O&S and/or life-cycle costs in acquisition decisions.

In practice, however, DoD cost-analysis activities in support of the acquisition process have emphasized acquisition over O&S costs. This is explained, at least in part and for the first two or three decades after the policies were first initiated, by the lack of a full capability within DoD to comprehensively and accurately capture O&S costs, understand the costs and their relationship to characteristics of defense systems, and develop the capability to estimate future O&S costs. Capturing O&S costs by weapon system is particularly challenging because O&S costs are typically generated over a long period in differing conditions that affect costs, are generated by many government and nongovernment organizations with differing cost accounting systems, are collected by disparate organizations, and are not the responsibility of any single organization.

DoD's ability to capture O&S costs by weapon system has evolved over time. In the 1980s, CAIG provided a cost element format for data collection (Recktenwalt, 1981), which began development of the Visibility and Management of Operating and Support Costs (VAMOSC) systems each military service uses today. The challenges remain formidable today, although the capability to capture O&S costs by system, and to use the data to estimate future costs, has improved.[8]

One indication of the relative emphasis on acquisition costs over O&S costs in OSD is that CAPE has had a separate O&S cost division only since FY 2013. Although there is a strong connection between design efforts early in development and subsequent O&S costs—and therefore some merit in having a cost analyst be familiar with the life cycle of a system and its costs—it is widely accepted in the cost-estimating community that the acquisition and O&S phases each require special skills and knowledge. Yet prior to the creation of CAPE in 2009, CAIG had just a single O&S cost analyst, while the services were placing greater emphasis on training and staffing O&S cost analysts.[9]

Given the long evolution in developing an O&S cost-estimating capability in DoD, and in OSD in particular, it is perhaps unsurprising that the acquisition policy goal of considering life-cycle costs in acquisition decisions has traditionally received little attention at milestone reviews. SMEs with long-term DoD experience reported during this research that, until recently, O&S was seldom discussed—and even less frequently briefed—to CAIG (now CAPE) or discussed at Defense Acquisition Board reviews. Researchers at the Institute for Defense Analysis noted that "interviewees reported that they could recall no instances when establishing or exceeding DTC goals was a topic of high-level deliberations. Review of several key decision documents of the period also revealed no discussion of DTC" (Kneece et al., 2014, p. iv). Regarding O&S costs, "[c]ontrolling longer-term O&S costs is substantially more difficult than nearer-term investment costs because of the uncertainty in O&S cost estimates, particularly early in the acquisition process," and "[i]t is difficult to motivate acquisition managers and contractors to maintain control over O&S costs" (Kneece et al., 2014, p. vii). We discuss these reasons for the difficulties in controlling O&S costs later in the report.

Organization of This Report

Chapter Two summarizes the legislation that created CAPE and that specified its initial duties and describes subsequent legislation that specified additional cost-analysis activities. Chapter Three describes CAPE O&S cost-estimating activities and historical

---

[8]   These challenges were more daunting before the development and widespread use of automated data-collection systems. In our assessment, it was not until the mid-1990s that DoD cost analysts typically had computerized access to a few years of O&S cost data by system.

[9]   For example, in 1986, the Naval Postgraduate School instituted an Operations Logistics curriculum within the Master of Science, Operations Research program. In 1990, the Commander, Naval Air Systems Command, established a separate integrated logistics support and O&S cost division inside the larger acquisition cost organization. Similarly, the Air Force Cost Analysis Agency established an O&S technical director to oversee weapon system O&S cost analyses and created a separate O&S cost division.

workload and the processes CAPE uses to accomplish these activities. Chapter Four describes the resources available in CAPE to accomplish its activities. Chapter Five provides our assessment of DoD's compliance with the requirements and guidance for CAPE O&S cost activities. Chapter Six presents a set of modest recommendations intended to enable CAPE to meet responsibilities added in laws through the FY 2017 NDAA. Chapter Seven provides a set of more-ambitious recommendations intended to position OSD to improve its oversight of O&S outcomes, including cost.

Appendix A provides an analysis of O&S cost estimates in the SARs that are submitted to Congress for MDAPs. Appendix B provides an overview of O&S funding captured in service VAMOSC systems. Appendix C provides a brief history of the organizational antecedents of CAPE.

Legal Requirements for CAPE O&S Cost Activities

This chapter focuses on legal and regulatory requirements related to CAPE O&S cost activities. Recent acquisition reform has taken the form of significant standalone legislation, such as WSARA (Pub. L. 111-23, 2009) and reforms incorporated into annual NDAAs. These regulations have, in turn, led to updated DoD directives, instructions, and guidance; the reorganization of regulatory structures; and changes in emphasis on certain aspects of acquisition.

We discuss key laws and regulations chronologically, beginning with WSARA in 2009, to give readers a sense of the increasing emphasis on O&S costs and related logistics outcomes.[1] We first describe the legislation that created CAPE and specified its initial duties, as well as other legislation that outlined additional cost analysis requirements.[2] We also discuss selected requirements in NDAAs that are applicable to the DoD components but that also affect CAPE workload. The chapter concludes by describing a key part of the FY 2018 NDAA—enacted in December 2017—the most recent attempt to affect O&S costs and improve logistics outcomes.

Establishment of CAPE: The Weapon Systems Acquisition Reform Act of 2009

WSARA established the office of CAPE, clarified the responsibilities of other offices in OSD, and established several new acquisition reform policies. WSARA is divided into three titles.

Title I of WSARA enacted changes to OSD's organization—most significantly, by establishing CAPE. Title I established a director of CAPE and two deputy directors, one for cost assessment (CA) and one for program evaluation. The CA deputy

---

[1] Section 846 of the FY 2017 NDAA repealed previous provisions in Title 10 related to major automated information systems (MAISs), so we have deleted references to those systems in the report except when describing historical CAPE workload.

[2] We mostly paraphrase language from legal and regulatory requirements, using words or short phrases from source documents. For ease of reading, we use quotation marks only for longer phrases or sentences.

director is responsible for cost estimation and cost analysis for acquisition programs. WSARA specifically tasks CAPE with issuing guidance regarding sustainability costs and full life-cycle management and requires CAPE to provide annual assessments of its cost activities. The following are CAPE's key cost estimation and cost analysis responsibilities for MDAPs, as outlined in Pub. L. 111-23, Section 101 (2009):

- conduct ICEs and ICAs in advance of
  - Milestone A, B, and C decisions
  - certification of any program that has a Nunn-McCurdy unit cost breach
  - a report of critical program changes, or on request
- participate in discussion of differences between CAPE's ICE and the military department's cost estimate
- review all cost estimates and cost analyses, including those conducted by the military departments and defense agencies.

Title II of WSARA outlined key reforms to acquisition policy, requiring DoD to consider trade-offs among cost, schedule, and performance objectives as part of the acquisition process. Title II requires CAPE input in analysis of alternative (AoA) decisionmaking, including being the primary lead to guide the conduct of and considerations involved in those analyses. The following are the key responsibilities, as described in Pub. L. 111-23, Section 201 (2009):

- The Director of CAPE will develop guidance for AoAs for joint military requirements validated by the Joint Requirements Oversight Council.
- The AoA guidance will require trade-offs among cost, schedule, and performance and assessment of whether the requirement can be achieved within cost and schedule goals.

Title III made additional minor provisions. Title III added procedures for reassessing MDAPs that experience critical cost growth in their program acquisition unit cost or procurement unit cost. As described in Pub. L. 111-23, Section 206 (2009), the Secretary of Defense, in consultation with CAPE, shall assess

- the estimated cost of completing the program of record
- the estimated cost of completing a reasonably modified program
- approximate costs of reasonable alternatives
- the necessity to cut funding for other programs because of the critical cost growth of the MDAP.

National Defense Authorization Act Provisions That Affect CAPE Workload

NDAAs since WSARA indicate congressional concern with product support, logistics outcomes, and O&S costs and have levied additional cost-estimating requirements on DoD and CAPE for MDAPs. The following summary of the NDAAs focuses on provisions that affect CAPE's responsibilities to estimate or assess O&S costs.

Section 864 of the FY 2011 NDAA (Pub. L. 111–383, 2011) required the Secretary of Defense to review DoD's acquisition guidance to determine, among other things, whether long-term sustainment of weapon systems is appropriately emphasized.

Section 832 of the FY 2012 NDAA (Pub. L. 112–81, 2011) required the Secretary of Defense to issue guidance to the military departments to do the following for MDAPs:

- periodically review O&S costs after initial operational capability (IOC) to identify and address causes of growth in O&S costs and develop strategies to reduce the costs
- update estimates of O&S costs periodically throughout the life cycle and retain the estimates and supporting documentation
- collect and retain data from operational and developmental testing and evaluation on R&M, and use the data to inform system design decisions, provide insight into sustainment costs, and inform estimates of O&S costs
- "ensure that sustainment factors are fully considered at key life cycle management decision points and that appropriate measures are taken to reduce operating and support costs by influencing system design early in development, developing sound sustainment strategies, and addressing key drivers of costs" (Pub. L. 112–81, Section 832(b)(7))
- "conduct an independent logistics assessment prior to key acquisition decision points (including milestone decisions) to identify features that are likely to drive future operating and support costs, changes to system design that could reduce such costs, and effective strategies for managing such costs" (Pub. L. 112–81, Section 832(b)(8))
- collect complete and accurate data in VAMOSC systems compliant with DoD standards and make the data available in a timely fashion
- establish standard requirements for the collection and reporting of contractor logistics support costs and develop contract clauses to ensure compliance.

It is noteworthy that DoD had formed a study team of government and industry personnel in 2008 to assess weapon system product support. The team recommended in 2009 that DoD issue policy to require the services to conduct independent logistics assessments, just as Pub. L. 112–81, 2011, did, and provide the results of the assessments to OSD in time for consideration before milestone decisions (Office of the Under

Secretary of Defense for Acquisition, Technology, and Logistics [OUSD(AT&L)], 2009, pp. 57–58). The law did not require the services to provide the results of the logistics assessments to OSD in time for consideration before milestone decisions, although 10 U.S.C. 2334 provides the authority for CAPE to obtain the data it needs, such as these assessments.

The FY 2012 NDAA gave CAPE the responsibility to keep a database of O&S estimates, supporting documentation, and actual O&S costs provided by the services. The responsibilities assigned to the military departments in the FY 2012 NDAA would subsequently affect responsibilities assigned to CAPE by the FY 2016 and FY 2017 NDAAs, which are summarized later.

Section 812 of the FY 2014 NDAA (Pub. L. 113–66, 2013) requires DoD to include risk and sensitivity analysis of estimates and schedule and technical risks in SARs and requires CAPE to review this information annually.

Sections 823 and 824 of the FY 2016 NDAA (Pub. L. 114-92, 2015) address requirements for Milestone A and B approvals for MDAPs, respectively. At Milestone A, the milestone decision authority must determine that an AoA in keeping with CAPE guidance has been conducted, that sustainment has been planned, and that a cost estimate of funds adequate to successfully execute the program through the life cycle has been submitted with the concurrence of CAPE.

Section 824 of the FY 2016 NDAA requires the Assistant Secretary of Defense for Research and Engineering to conduct an independent technical review of MDAPs at Milestone B to determine "that the technology in the program has been demonstrated in a relevant environment."[3]

Furthermore, at Milestone B, the milestone decision authority must determine that life-cycle sustainment planning has found and assessed the sustainment costs of the program and the costs of alternatives throughout the life cycle, that the costs are sensible and have been estimated accurately, and that core logistics workloads and capabilities have been estimated.

Section 807 of the FY 2017 NDAA (Pub. L. 114–328, 2016) requires cost, schedule, and performance goals for milestone decision authority approval of MDAPs at Milestones A, B, and C. Cost goals must be set for both procurement unit cost and sustainment cost.

Section 842 of the FY 2017 NDAA requires that CAPE complete or approve an ICE prior to Milestones A and B, and C:

- The ICE at Milestone A must include sensitivity analysis of key cost drivers affecting life-cycle costs.

---

[3] DoD uses a scale from 1 through 9 of technology readiness levels, with higher levels indicating greater technological maturity. The criterion "demonstrated in a relevant environment" corresponds most closely with level 6 "[s]ystem/subsystem model or prototype demonstration in a relevant environment." See Assistant Secretary of Defense for Research and Engineering (2011).

- The ICE at Milestones B and C must include life-cycle costs and identify and evaluate alternative courses of action that may reduce cost and risk.

To increase transparency of DoD acquisition decisions to Congress, Section 808 of the FY 2017 NDAA requires the milestone decision authority to provide reports to Congress after Milestone A, B, and C approvals. These reports include summaries of the cost, schedule, and risk information approved as part of the milestone decision.

To improve life-cycle cost control of MDAPs, Section 849 of the FY 2017 NDAA added a new section, 2441, Sustainment Reviews, to 10 U.S.C. The new law requires the secretary of each military department to conduct sustainment reviews of MDAPs five years after IOC and throughout the life cycle of the program. The reviews are to include

- an ICE for the remainder of the life cycle
- a comparison of actual and budgeted costs
- a comparison of planned and achieved reliability
- "an analysis of the most cost-effective source of repairs and maintenance" (10 U.S.C. 2441 (b)(4))
- "an evaluation of the cost of consumables and depot-level repairables" (10 U.S.C. 2441 (b)(5))
- "an evaluation of the costs of information technology, networks, computer hardware, and software maintenance and upgrades" (10 U.S.C. 2441 (b)(6))
- an assessment of planned and actual fuel efficiencies
- a comparison of estimated and actual manpower requirements.

It is again noteworthy that the law mandating post-IOC reviews is similar to a recommendation from the DoD Weapon System Product Support Team for post-IOC reviews led by the Logistics and Materiel Readiness office in OSD and the service responsible for life-cycle management (OUSD[AT&L], 2009). However, unlike the DoD recommendation, the law does not include OSD in the review process.

Further Indications of Congressional Concern: The Fiscal Year 2018 NDAA

The FY 2018 NDAA enacted in December 2017 added a new code section, 10 U.S.C. 2443, Sustainment Factors in Weapon System Design. The new section

- requires R&M as "attributes of the key performance parameter on sustainment during the development of capabilities requirements"
- requires the program manager to include requirements for R&M engineering activities and design specifications in contracts for EMD and production. If the

program manager determines that R&M should not be a contract requirement, "the program manager shall document in writing the justification for the decision"

- requires the Secretary of Defense to ensure that sustainment factors, including R&M, are emphasized in the source-selection process and encourage consideration of objective R&M criteria
- authorizes offering incentive fees to contractors that exceed the design specification requirements for R&M and require the use of recovery options for failure to meet the design specification requirements for R&M.

The House Armed Services Committee report on the House version of the FY 2018 NDAA explained the intent of the new section of law:

> The committee notes that the design of a major weapon system directly affects its life-cycle sustainment activities and consequently drives its O&S costs. Elements of sustainment that are highly dependent on the system design, namely R&M, are easier and less costly to address during the development of an MDAP than after a weapon system is fielded. Therefore, the committee believes the Department should emphasize R&M in early engineering decisions. (Thornberry, 2017, p. 164)

Although the new law to require sustainment factors in weapon system design has no explicit additional requirement for independent cost estimates, the clear intention is to affect O&S costs. In addition, the requirements for R&M engineering activities and the provisions for financial penalties and awards tied to achieving the R&M metrics will inevitably affect broader product-support costs incurred from development through O&S.

Summary

This chapter examined key laws and regulations relevant to CAPE, finding an increasing emphasis over time on O&S costs and related logistics outcomes. In 2009, WSARA created the office of CAPE and added a new section (Section 2334) to 10 U.S.C. that specifies that "[t]he Director of Cost Assessment and Program Evaluation shall ensure that the cost estimation and cost analysis processes of the Department of Defense provide accurate information and realistic estimates of cost for the acquisition programs of the Department of Defense."

Since passing this law, Congress has signaled its concern about weapon system logistics and O&S cost outcomes with additional legislation that mandates cost-analysis activities intended to improve these outcomes. The most significant changes were legislated in the FY 2012, FY 2016, and FY 2017 NDAAs:

- The FY 2012 NDAA established new requirements for the services to conduct independent logistics assessments at milestone reviews of major programs; to consider sustainment at key life-cycle management decision points; to take measures to reduce O&S costs by influencing system design early in development; and to review O&S costs, update O&S cost estimates, and address O&S cost growth in reviews after system IOC. The law specified that these additional requirements are the responsibility of the military departments, not CAPE or OSD.

- The FY 2016 NDAA required sustainment planning and estimates of sustainment costs for MDAPs. The Director of CAPE must ensure, at Milestone A, that the level of resources required to develop, procure, and sustain the program is sufficient to execute the program. At Milestone B, the milestone decision authority must determine that life-cycle sustainment planning has identified and evaluated relevant sustainment costs throughout the life cycle, and the costs of any alternatives, and that the costs are reasonable and have been accurately estimated.

- The FY 2017 NDAA added requirements for analysis done in ICEs. The ICE at Milestone A must include sensitivity analyses of key cost drivers that affect life-cycle costs. The ICE at Milestones B and C must include life-cycle costs and identify and evaluate alternative courses of action that may reduce cost and risk. CAPE must conduct or approve these analyses.

- Finally, the FY 2018 NDAA requires incorporation of R&M metrics as a key performance parameter for sustainment when determining capability requirements. Program managers must include R&M metrics in development and procurement contracts or provide written justification. The law gives DoD the authority to write contract provisions to hold weapon system contractors financially accountable for meeting the metrics. The FY 2018 NDAA clearly indicates continuing congressional concern with improving weapon system R&M and reducing O&S costs.

The FY 2012, FY 2016, and FY 2017 NDAAs specify a great deal of planning, analysis, and cost estimation of product support activities for MDAPs that was not previously required by law. Crucially for this report, the law that created CAPE requires it to conduct or approve the cost estimates of these activities. This expansion from only estimating O&S costs to including consideration of product support activities broadens CAPE's responsibilities.

CHAPTER THREE

CAPE Operating and Support Cost-Estimating Activities, Workload, and Processes

This chapter describes CAPE O&S cost-estimating activities, workload, and processes; Chapter Four focuses on CAPE resources available to carry out these activities. The discussion in these two chapters will support our assessment of CAPE's compliance with recent laws regarding O&S cost estimation (Chapter Four).

To identify CAPE O&S cost-estimating activities and determine what drives them, we conducted structured interviews with each of the O&S cost analysts in CAPE. We conducted unstructured interviews with CAPE analysts with specialized duties for tracking legal requirements for CAPE cost activities and expertise in estimating methodologies for product-support costs incurred during acquisition. We conducted the same structured interviews with groups of the O&S cost analysts and/or their managers from each of the service's independent cost-estimating organizations to learn similarities and differences between their activities and methodologies and CAPE's. The primary purpose of these interviews was to supplement knowledge of cost-estimating procedures we had from written guidance and our own experience in doing independent cost estimates, with a secondary purpose of discovering practices the services used that CAPE could adopt. We analyzed the data through a tabular comparison of results across respondents, focusing on the time, personnel, data sources, and methodologies used to conduct O&S estimates. We also reviewed the *Annual Report on Cost Assessment Activities* for FYs 2010–2016 to understand and measure CAPE workload (Director, CAPE, 2011–2017).

Activities

The main drivers of CAPE O&S cost-estimating activities are milestone decision-review events for MDAPs as the programs proceed through the acquisition process.[1]

---

[1] Although MAIS and pre-MAIS milestones and reviews should have also driven O&S cost activities, there are too few O&S cost analysts to support these events, so cost analysis for MAIS and pre-MAIS milestones and reviews have been conducted by other CAPE analysts.

The blue columns in Figure 3.1 show the annual count of events that drive O&S cost-estimating and review activities. The red columns in the figure show the annual count of events that drive acquisition cost-estimating and review activities.

The events that drive acquisition cost-analysis activities in CAPE include the same milestone decision reviews that drive O&S cost activities plus multiyear procurement awards and Nunn-McCurdy breaches. Separate divisions of cost analysts in CAPE perform O&S and acquisition estimates, so we display each type of activity separately.

In addition to events associated with MDAPs moving through the acquisition process, another driver of O&S cost-estimating activity in CAPE is support of special studies mandated in legislation or requested by DoD offices. An example of a special study is the CAPE ICE performed in 2015 for a Missile Defense Agency program to assess several programmatic changes. Congress requested the ICE in the FY 2015 NDAA. The number and scope of special studies are difficult to predict and difficult to assess as drivers of workload because the nature and amount work can vary considerably from one project to another. The special studies are not shown in the counts in Figure 3.1.

Workload and Processes

Most of the cost-analysis workload in CAPE involves performing or reviewing ICEs of MDAPs at reviews during key milestones in the acquisition process. A CAPE ICE for

Figure 3.1
CAPE Annual Cost-Estimating Activities, FYs 2010–2016

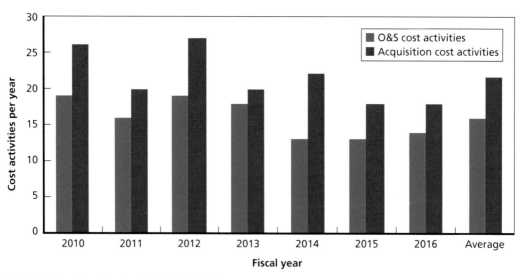

SOURCE: Based on Director, CAPE, 2011–2017.
RAND RR2527-3.1

an MDAP must begin at least 180 days prior to any acquisition milestone decision.[2] CAPE O&S cost analysts typically do not work exclusively on one MDAP at a time because there are, inevitably, periods of waiting associated with scheduling meetings with SMEs, obtaining documents and data, and the like. Thus, cost analysts typically work on at least one other project while conducting an ICE of any given MDAP. CAPE analysts said that it takes them about six months to complete an estimate, although the actual length of time depends on the type of review being conducted, data availability, and other factors.

An important starting point for an estimate is the cost analysis requirements description (CARD) for the MDAP. The CARD provides a detailed description of the program, which establishes a common understanding and is used in preparing estimates by cost analysts in CAPE, independent service cost organizations, and program offices. The CARD must be signed by the program manager and program executive officer and is provided, at least in draft form, to CAPE no later than 180 days before the milestone decision.

The CARDs we read for some of the MDAPs we assessed and describe in Appendix A contain sections on the following (DoD was revising the reporting requirements for CARDs when this report was written):

1. system overview
   1.1. system characterization and description
   1.2. system characteristics; technical and physical characteristics
   1.3. reliability, maintainability, availability
   1.4. embedded security
   1.5. predecessor system
2. risk
3. operational concept
4. quantity
5. manpower
6. activity rates
7. milestone schedule
8. acquisition plan or strategy
9. system development plan
10. facilities requirements
11. track to the prior CARD
12. cost and software data reporting plan.

---

[2] Timelines and events for cost-analysis activities are provided in DoDI 5000.73 (2015). The 180-day requirement is also specified in DoDI 5000.02 (2017) .

A second document O&S cost analysts use, although much less valuable to analysts than the CARD, is the life-cycle sustainment plan (LCSP) for a program. Program managers are responsible for developing the plan starting at Milestone A. The LCSP evolves over time. At Milestone A, the LCSP begins to develop sustainment metrics to influence design and the product-support strategy and to reduce O&S costs on actions prior to system development. At Milestone B, the LCSP should include metrics for materiel reliability, O&S cost, mean downtime, and other sustainment measures (DoDI 5000.02, 2017, Encl. 6).

The approach for CAPE estimates of MDAPs is typically a modified analogy to similar programs, with complexity adjustments based on technical reviews of the programs. Programs without close analogs, such as those with significantly new technology, require more interaction with the contractor and are typically more difficult to estimate. Our analysis of estimates of unit O&S costs reported in SARs, presented in Appendix A, confirms this.

Based on our interviews with CAPE O&S cost analysts, they spend about 80 percent of their time collecting data from the program manager, weapon system contractor, resource sponsor, and end users who will operate and support the system. O&S cost analysts use data on antecedent systems from such sources as the official O&S cost database of each service,[3] program office predictions of performance, and prime contractor materials list.

These data offer insights into how the program under review differs from its antecedent in ways that affect cost, including differences in capability, technical complexity, reliability, maintainability, sustainment approach, and concept of operations.

The historical norm described earlier is based on available documentation for the program of record. However, depending on how DoD interprets and implements new requirements for sustainment cost estimates in the FY 2016 and FY 2017 NDAAs, the legislation could expand CAPE duties for cost estimates to include alternative courses of action in addition to the program of record and assessing and estimating elements of sustainment costs in greater detail than is traditional. The expanded duties could require identifying changes, perhaps proposed by the program office or resource sponsor, that might be more cost-effective than the program of record and estimating the cost of the alternatives. It is not clear how DoD will interpret and implement this requirement. The requirements in the FY 2016 and FY 2017 NDAAs could add considerably to the workload of O&S cost analysts if the changes to the program of record are interpreted to include redesigning some equipment features. However, if they are interpreted more narrowly, as programmatic changes in quantities, production rate, flying hours, etc., the impact on workload could be small. Chapter Five provides our assessment of CAPE's capability to implement the new requirements.

---

[3] Each service is required by law to have a data system that collects and retains O&S costs for its weapon systems. These data systems are generically referred to as *VAMOSC systems.*

Summary

- The main driver of CAPE O&S cost-estimating activities is milestone decision review events for MDAPs as the programs proceed through the acquisition process.
- Most of CAPE's cost-analysis workload involves conducting or reviewing ICEs of MDAPs at milestone reviews. The key document used in this work is the CARD, which provides the technical and programmatic requirements needed for the cost analysis.
- The approach for CAPE estimates of MDAPs is typically a modified analogy to similar programs, with complexity adjustments based on technical reviews of the programs. Programs without close analogs, such as those with significantly new technology, require more coordination with the contractor and are typically more difficult to estimate.
- A cost analyst spends about 80 percent of his or her time collecting data from the program manager, resource sponsor, and end users who will operate and support the system. O&S cost analysts also use data on antecedent systems from such sources as the official O&S cost database for each service, program office predictions of performance, and the prime contractor's list of materials.
- Another driver of O&S cost-estimating activity in CAPE is support of special studies mandated in legislation or requested by DoD. Unlike cost analyses to support milestone decisions, the number of special studies each year and their workload is highly variable.

CHAPTER FOUR

CAPE Resources for O&S Cost Activities

This chapter describes the resources available to CAPE for O&S cost-estimating activities. Personnel and data are the two key resources CAPE needs to meet requirements for estimating O&S costs. We also describe several issues with data that can affect analysts' ability to conduct cost estimates.

Personnel

The first key resource in CAPE is personnel. CAPE's cost-estimating personnel include a deputy director and assistant, four analysis divisions, and a center for managing cost data. Table 4.1 shows the numbers of personnel assigned to cost assessment organizations in 2017.

The basic estimating responsibilities of the four analysis divisions and the Cost Assessment Data Enterprise/Defense Cost and Resource Center are as follows:

Table 4.1
Personnel Assigned to Cost Assessment Organizations in 2017

Organization in Cost Assessment	Government (number)	Contractor (number)	Total
Cost Assessment deputy director and staff	2	0	2
Operating and Support Cost Analysis Division	5	1	6
Economic and Manpower Analysis Division	9	1	10
Weapons Systems Cost Analysis Division	10	0	10
Advanced Systems Cost Analysis Division	15	1	16
Cost Assessment Data Enterprise/Defense Cost and Resource Center	3	78	81
Total	44	81	125

SOURCE: Based on information provided by OSD/Cost Assessment.

- The Operating and Support Cost Analysis Division conducts ICEs and ICAs for the O&S phase of an MDAP.
- The Economic and Manpower Analysis Division analyzes broader issues, such as health care costs, personnel compensation, macroeconomic forecasting, and similar issues not focused on specific MDAPs.
- The Weapons Systems Cost Analysis Division and the Advanced Systems Cost Analysis Division conduct ICEs and ICAs for the development and production phases of an MDAP.
- The Cost Assessment Data Enterprise/Defense Cost and Resource Center collects MDAP cost, schedule, technical, and programmatic data and makes the data available to authorized government analysts.

Data

Data are the second key resource needed to enable O&S cost estimating. The issues described below regarding data apply both to independent cost analysts in the service cost-estimating organizations and to CAPE analysts. These cost analysts report needing several types of data to do their work, including

- the costs of antecedent systems and subsystems or components and (if available) of the system for which costs are being estimated
- technical descriptions or characteristics of the system for which costs are being estimated, as well as for the antecedent system
- descriptions or similar information about the content of repair or maintenance activities performed
- usage rates for the system (such as average flying hours per aircraft per year or average steaming hours per ship per year)
- the expected reliability of major subsystems or components of the system
- requirements for scheduled maintenance
- the approach to sustainment.

Table 4.2 summarizes the primary sources of these data for cost analysts. The sources shown in columns 2 through 5—the CARD, LCSP, cost and software data report (CSDR), and VAMOSC systems—are formal and institutionally supported. DoD has invested and continues to invest considerable time and effort in them. Technical descriptions and characteristics and projected usage rates are available in the CARD, and the LCSP describes the sustainment approach. DoD guidance requires MDAP program managers to prepare and provide these documents for review in the acquisition milestone process, as discussed in Chapter Three.

Table 4.2
Primary Sources of O&S Data for Cost Analysts

Data Need	CARD	LCSP	CSDR	VAMOSC Systems	Other
Technical descriptions and characteristics	X				SMEs
Cost			X	X	
Usage rates	X			X	
Reliability or expected reliability of major subsystems or components					SMEs, databases
Requirements for scheduled maintenance					SMEs, databases
Content of repair or maintenance activities					SMEs, databases
Sustainment approach		X			

NOTE: *Databases* refers to the logistics and financial management databases controlled by services or service logistics and comptroller organizations.

Analysts obtain cost data from two main sources: CSDRs and VAMOSC systems. CSDRs contain actual costs and related information supplied by DoD contractors.[1] CAPE analysts use the reports on legacy systems as the basis for estimating the costs of future analogous systems. Reports are required for contracts valued at more than $50 million for MDAPs after Milestone A approval. DoD collects and retains the reports in the Defense Cost and Resource Center.

O&S cost analysts also rely heavily on data from VAMOSC systems. Each military service funds, manages, and collects and reports data in its own VAMOSC system. The Army's VAMOSC system is called the Operating and Support Management Information System. The Navy and Marine Corps Navy system is called VAMOSC (which we refer to as *Navy VAMOSC*). The Air Force's VAMOSC system is called Air Force Total Ownership Cost (AFTOC).[2] These reporting systems collect the following costs by weapon system:

- unit-level military and civilian personnel by billet and pay grade (the Army's Operating and Support Management Information System does not have these data)
- unit-level consumption, including energy, munitions, and administration
- weapon system maintenance, overhaul, and modification.

---

[1]   CSDR content and requirements for submitting CSDRs are described in DoD Manual 5000.04 (2018).

[2]   We used data from Navy VAMOSC and AFTOC in conducting analyses in this project. The URLs are listed in the references, although they are not available to the general public.

VAMOSC systems report costs; usage data, such as aircraft sorties and flying hours, ship steaming hours, and vehicle miles; and programmatic data, such as number of units of the weapon system in the inventory at a given time. The standard cost element structure calls for including weapon system sustaining support, including system engineering, program management, and technical data costs. However, not all VAMOSC systems are equally able to collect these data by weapon system.

Some data O&S cost analysts need are not available from the institutionalized sources just described. Analysts indicate that most of their time is spent gathering such additional data from other sources, as described in Chapter Three.

Issues with Data

O&S cost analysts encounter several data-sufficiency issues while conducting ICEs. We describe these issues by data source.

CARDs and LCSPs

We reviewed CARDs and LCSPs for several MDAPs to determine the extent to which the data they contain are useful for informing O&S cost estimates. The CARDs and LCSPs were for programs projected in SARs to have among the largest life-cycle O&S costs in DoD, including the F-35, V-22, LCS, and F-22. We also reviewed program office estimates or ICEs prepared for the programs early in development to determine how consistent the assumptions underlying the estimates are with the data provided in the CARD. The documents contain restricted information that cannot be cited in this unrestricted report, so we will provide a general assessment of their usefulness.

We found that the information provided in the CARDs is more helpful for estimating acquisition—rather than O&S—costs, although the CARDs did provide system-level goals for R&M metrics. The CARDs also provided projections of required numbers of crew, maintenance, and other unit-level personnel.

The LCSPs examined did not provide much useful O&S information beyond what the CARDs already supply.[3] RAND research team members with experience as former government cost estimators recalled that the results of detailed logistics-support analyses would, at best, be available late in a development program and by special request to a program office or associated organization. Only late in a development program is sufficient information available to perform such analyses; even then, substantial uncertainty remains about reliability and other key factors affecting O&S costs and other aspects of logistics support. However, information from supportability

---

[3]   One might expect that LCSPs would include information from logistics support analyses, which used to be mandated for DoD programs in Military Standard (MIL-STD) 1388-1, 1993, but that was not the case.

analyses during development, such as maintenance task analysis and level-of-repair analysis, could be useful to cost analysts and could be made available to them.

We reviewed estimates of O&S costs prepared during development for the F-35, V-22, and F-22 programs. The estimates reflected the personnel requirements and R&M metrics shown in the CARDs for the respective programs. Although the F-22 and V-22 are more mature than the F-35, all three programs had been fielded as of 2017 and have generated data that can be compared with the CARD inputs and with the cost estimates based on the inputs. With the benefit of hindsight, we identified the following problems with the inputs and resulting cost estimates:

- The number of unit-level personnel needed to support the system was understated in the CARDs for two programs and was well below the actual number of personnel assigned.
- R&M metrics were provided in the CARD at the system level but not at a low enough level to be helpful in estimating costs of subsystems or components, such as engines, radars, or other equipment subject to costly maintenance. In addition, the system-level metrics were overstated compared with actual experience as reported in DOT&E reports and/or SARs.
- As an input to estimating maintenance costs, information on expected reliability or *frequency* of repair is only some of what is required. Information is also needed on the *cost* to repair. In the absence of actual operational data early in development, the estimates we examined used repair costs for antecedent aircraft adjusted for the expected improved reliability reported in the CARD for some elements, thereby combining the most optimistic assumptions regarding reliability and cost. Our analysis of the procurement and repair costs for aviation components shows an increase above the rate of inflation over many years.

Other Data Sources

DoD collects cost and programmatic data on weapon systems that would be useful for cost analysis but that are not routinely provided to cost analysts. One example is the annual Depot Maintenance Cost System data provided to OUSD(AT&L). The data cover DoD depot maintenance facilities and private-sector facilities; are reported by weapon system; and provide detail on labor hours, labor, material, and other categories of cost and on the nature of the work performed (DoD 7000.14R, 2016).

Another example is the supply system inventory reporting DoD components provide to OUSD(AT&L). The data include the level of supply inventory associated with a weapon system program and purchased by a program manager for initial or sustainment program support, whether managed by the government or a contractor. The information can be used to assess the ability of the supply inventory to meet stated requirements (DoD Manual 4140.01, 2017).

Reliability measures require failure data, along with population size and utilization. Early indications of these measures are captured during operational test and evaluation by DoD test organizations.

Historical costs for product support elements—such as peculiar support equipment, peculiar training equipment, publications and technical data, initial spares, and information on the adequacy of this support relative to requirements for it—are especially important to the estimation of product support for future systems. As noted, such information is lacking in CSDRs.

The cost and programmatic information on product support elements, such as support equipment, training devices, and initial spares for organically supported legacy programs, should be available from the government facilities that provide the support. However, it is extremely difficult for cost analysts to find this information, and we are not aware of a repository of such information in any DoD organization.

CSDRs

One data issue with CSDRs is a gap in data on acquisition costs for the 1990s and early 2000s. At the time, there was a belief among some in DoD that DoD requirements for cost reporting were overly expensive and burdensome to contractors. This feeling, combined with the acquisition reform goals of that era, caused far fewer cost reports to be submitted than in prior and later periods. For example, most Navy programs had waivers until relatively recently, and there were no reports for CG-47, FFG, SSBNs, or DDG-51–class ships through DDG-112.[4] The acquisition costs for weapon systems are relevant to analysts who estimate O&S costs because product support elements that affect O&S costs—such as peculiar support equipment, peculiar training equipment, publications and technical data, and initial spares—are funded with development or procurement funds (or both).

A second issue is that DoD's Cost Assessment Data Enterprise, which collects and stores CSDRs and other data, has collected CSDRs for sustainment contracts only since 2010. This means that there is little visibility into sustainment costs provided by contractor logistics support for most programs before that time. VAMOSC systems have, until recent years, shown a single sum for the entirety of the sustainment contract value, with no breakdown by the standard cost elements used for DoD O&S costs. Today, CSDRs for sustainment contracts report the costs by element, and VAMOSC systems are increasingly able to report the contractor logistics support costs by element. Figure 4.1 illustrates both the total number of programs that have submitted CSDRs and the subset of these that are sustainment programs. As shown in the figure, DoD is rapidly adding to its repository of sustainment CDSRs.

A third issue with CSDRs is that they often omit programmatic information— such as quantities of units bought or services provided—that, combined with costs,

---

4   Cost Assessment Data Enterprise SME, communication with the authors, April 10, 2017.

Figure 4.1
Number of Programs Providing Cost Reports, Total and Sustainment

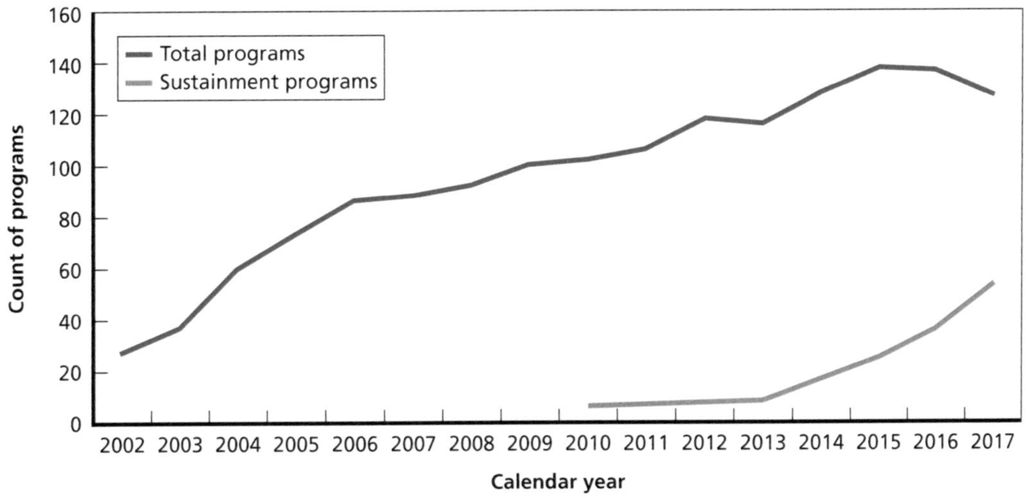

SOURCE: Based on data from DoD Cost Assessment Data Enterprise.
NOTE: 2017 reflects partial year data.
RAND RR2527-4.1

provides meaning and context to the data. For CSDRs of acquisition contracts, programmatic information relevant to sustainment includes such information as quantities of unique support equipment, training devices, and initial spares. Programmatic information in CSDRs for sustainment contracts would indicate, for example, the number of scheduled or unscheduled repairs or other maintenance activities. Our previous experience with CSDRs is that they do not provide programmatic information.

To assess the programmatic information reported on current programs, we checked a small sample of CSDRs for acquisition contracts for the following five recently developed or currently developing programs: F-35, Joint Light Tactical Vehicle, KC-46, LCS, and P-8. CSDRs have a column to report data to date and another column to report data estimated at completion of the contract. For elements relevant to sustainment—such as training, data, peculiar support equipment, common support equipment, site activation, and industrial facilities—CSDRs rarely reported actual or estimates-at-completion-of quantities. In lieu of a better method for estimating the cost of these elements—which are part of sustainment costs and which determine future O&S costs—cost analysts typically estimate them as a factor of the cost of the prime mission product. Unfortunately, there is so much variance in the historical factors, and so little insight into the adequacy of the historical funding, that using factors as an estimating methodology provides little assurance of adequate funding for these elements. CAPE's Cost Assessment Data Enterprise is working on a requirement to include more programmatic information in sustainment CSDRs.

A fourth data issue is that CSDRs and other cost and technical data are seldom available for the commercial portions of commercial-derivative systems, such as the Air Force's KC-46 tanker or the Navy's P-8 patrol aircraft. These aircraft are based on commercial aircraft platforms (the KC-46 is based on the Boeing 767, the P-8A on the Boeing 737) but involve extensive modifications to address military requirements. Vendors are required to provide cost and related technical information for products developed with government funding but not for products, including subsystems or components of weapon systems, developed wholly or in part with the vendor's funding.

Legislation in two recent NDAAs illuminate issues regarding data for commercial or commercial-derivative items and DoD's ability to request such data. Section 831 of the FY 2013 NDAA (Pub. L. 112–239, 2013) requires OSD to issue guidance on data requests other than certified cost or pricing data necessary to determine the reasonableness of the price of the contract when buying commercial items. The NDAA stipulates that the guidance shall "provide that no additional cost information may be required by the Department of Defense in any case in which there are sufficient non-Government sales to establish reasonableness of price."

The government's rights to technical data are addressed in Section 813 of the FY 2016 NDAA (Pub. L. 114–92, 2015). Entitled "Rights in Technical Data," the section added language to 10 U.S.C. 2321(f) to clarify that a contractor may restrict the release of technical data for a commercial subsystem or component of an MDAP that was developed exclusively at private expense. Section 813 of the FY 2016 NDAA also requires the Secretary of Defense to establish a government-industry advisory panel to review the law regarding rights in technical data, proprietary data restrictions, and relevant regulations.

Industry representatives have voiced concerns that government assertions of rights to cost and technical data can discourage industry from doing business with DoD, thus stifling both innovation and the introduction of technology to DoD systems. Policymakers continue to try to balance the need for information for DoD to acquire and sustain commercial-derivative systems with the concerns of the industrial base.

VAMOSC Systems

The services' VAMOSC systems have greatly enhanced cost analysts' access to weapon system–oriented cost and programmatic information. This information is collected and reported in numerous data systems in each service, and the VAMOSC systems integrate and organize the information by weapon system in the standard O&S cost-element structure. Cost and programmatic data are the lifeblood of cost analysis. The accurate accounting and timely availability of the costs for currently operating weapon systems are essential in determining the costs for these systems, forecasting their future costs, and estimating the costs of analogous weapon systems. The VAMOSC systems are therefore extraordinarily important to O&S cost analysts.

Although CAPE is responsible for providing guidance regarding the services' VAMOSC systems, responsibility for funding and populating the data in the systems lies with each service. The VAMOSC systems are managed by service headquarters organizations, and contractors provide expertise in database management. The systems have thus been especially vulnerable to budget cuts that target headquarters activities and contract services. Table 4.3 shows total budgeted resources for the three VAMOSC systems as of summer 2017.

The VAMOSC systems provide annual operating costs and basic programmatic information (such as the number of weapon systems in the inventory and their usage). However, these systems do not provide sufficient programmatic or performance data to determine cost-rate factors (such as cost per engine, aircraft, ship, or vehicle overhaul) necessary for cost analysis and especially for forecasting future O&S costs. With depot maintenance costs, for example, the total annual cost is generally reported without reporting the corresponding number of completed events or their work scope. Effective use of the cost data requires programmatic or performance data that reside with the military services but that are often not reported in VAMOSC systems or elsewhere readily accessible to cost analysts.

Programs that use contractor logistics support provide less cost detail and less programmatic and performance data than organically supported systems. Increased use of contractor logistics support means that less information is available on a larger number of fleets and a larger proportion of weapon system O&S costs than in the past.

Addressing these challenges will require adequate funding of VAMOSC systems and firm direction from OSD to require the services to provide the programmatic and performance data that resides in service depots and similar organizations.

Authority

The law requires CAPE to have sufficient personnel to perform its independent cost estimation and cost analysis duties and empowers CAPE to obtain cost and programmatic data and to access expertise that resides in other DoD organizations. The necessary authority is granted in 10 U.S.C. 2334, as previously noted, and in DoDD 5105.84 (2012), which states that other OSD principal staff assistants and the heads of DoD components are to ensure that CAPE has timely access to any data it requires to perform its duties. However, actual experience is mixed. CAPE analysts are sometimes

Table 4.3
Budget Resources for VAMOSC Systems, Then-Year $ Millions

	FY 2016	FY 2017	FY 2018	FY 2019	FY 2020	FY 2021	FY 2022
Budget amount	12.7	11.3	10.7	9.8	9.8	11.9	11.9

SOURCE: Based on data from CAPE.

unable to obtain subject-matter expertise and data from other organizations. Given the relatively short amount of time CAPE O&S analysts are involved with a program, timely access is critical to informing their analyses, and the analysts cannot always overcome the access issues to obtain the needed information.

Summary

- Personnel and data are the two key resources needed to meet requirements for estimating O&S costs. CAPE's O&S cost-estimating personnel include a deputy director, an assistant, and four analysts, supported by a center for managing cost data. O&S cost analysts need several types of data for their work, including data on the costs of antecedent systems, technical descriptions of the MDAP being estimated, information about repair or maintenance activities, usage rates, reliability, scheduled maintenance, and the sustainment approach.
- Analysts obtain cost data from two main sources: CSDRs and VAMOSC systems. CSDRs contain actual costs and related information supplied by DoD contractors. Each military service funds, manages, collects, and reports data in its own VAMOSC system.
- O&S cost analysts encounter several data-sufficiency issues while conducting ICEs.
- The information provided in CARDs is more helpful for estimating acquisition—rather than O&S—costs, although the CARDs do provide system-level goals for R&M metrics.
- The LCSPs examined did not provide much useful O&S information beyond what the CARDs already supply.
- There are several problems with CDSRs, including a gap in data on acquisition costs for the 1990s and early 2000s, lack of sustainment data prior to 2010, insufficient programmatic information, and insufficient data on commercial vendor costs.
- The services' VAMOSC systems provide data on annual operating costs and basic programmatic information; however, these systems do not provide sufficient information to determine cost rate factors for all O&S cost elements.
- Fewer cost details and fewer programmatic and performance data are available for programs that use contractor logistics support.
- Other data that exist but that are not currently made available to CAPE could be useful for O&S cost analyses. These data include annual Depot Maintenance Cost System data, supply system inventory reporting, and detailed cost and programmatic information on product support elements for legacy systems. CAPE has the authority to obtain these data.

CHAPTER FIVE

Assessment of CAPE's Compliance with Recent Laws Regarding O&S Cost-Estimating Activities

This chapter draws from the findings presented in the previous two chapters to inform our assessment of CAPE's compliance with recent laws concerning O&S cost-estimating activities. We first explain the criteria used in the assessment and then provide the results.

Criteria for Assessing Compliance

We considered two different criteria for assessing compliance. The first comprises *the time, effort, data, and organizational resources*—described in Chapter Three—traditionally used in CAPE ICEs and ICAs. The second is more demanding and recognizes the recent expansion in scope of CAPE's duties from a more traditional focus on O&S costs to include product support and *the intention behind the laws to motivate program managers and military departments to acquire affordable and reliable systems* as enacted in the FY 2016, 2017, and 2018 NDAAs:

- Sections 823 and 824 of the FY 2016 NDAA (Pub. L. 114–92, 2015) require planning for sustainment to be addressed and life-cycle costs to be estimated for Milestone A and B approval. It seems clear that Congress intended for sustainment planning to be accomplished early and funded adequately.
- Section 807 of the FY 2017 NDAA (Pub. L. 114–328, 2016) requires OSD to set procurement unit cost and sustainment cost goals for MDAPs. Section 842 of the FY 2017 NDAA requires CAPE to conduct or approve ICEs at milestone reviews that identify and evaluate cost and risk drivers and alternative courses of action that may reduce cost and risk.

Thus, the law acknowledges the connection between sustainment programmatics and cost and requires OSD, and CAPE in particular, to play a role in evaluating both.

The FY 2018 NDAA signed into law in December 2017 further telegraphs the intent of Congress to motivate improved R&M in DoD weapon systems. The law also indicates that Congress recognizes that system engineering decisions made early in the acquisition process affect logistics outcomes and sustainment costs.

Our second criterion in assessing compliance with recent laws thus requires a higher level of CAPE resources and involvement in oversight of MDAPs. The higher level of resources and involvement would allow cost estimates for MDAPs that are sufficiently accurate and detailed to ensure the adequacy of product support activities during execution. Such estimates would better inform decisionmakers during acquisition milestone reviews and help provide and ensure incentives for successful sustainment outcomes. This second, more-rigorous criterion informs the recommendations provided in the final chapter of this report.

Assessment Findings

Our assessment of compliance is based on the level of resources traditionally used in CAPE ICEs and ICAs. The degree of compliance is organized by the main requirements pertaining to CAPE O&S cost-estimating duties in 10 U.S.C. effective as of October 2017. The requirements are organized by section and are summarized in Table 5.1.

The table shows that our team found room for some improvement across all areas of CAPE compliance with laws regarding O&S cost-estimating activities. Two areas were assessed as being noncompliant: (1) reviewing the cost and associated information in SARs and (2) assessing life-cycle cost estimates associated with post-IOC sustainment reviews.

The FY 2014 NDAA amended 10 U.S.C. 2334(a) to require CAPE to annually review the cost and associated information in SARs (Pub. L. 113–66, Section 812, 2013). This analysis of SARs is not currently done in OSD. The Office of the Assistant Secretary of Defense for Logistics and Materiel Readiness currently reviews all SARs for consistency and coherency, but the review does not entail the substantive level of analysis the law requires CAPE to perform. In 2017, DoD had 30 days for the current review, from submission of the President's Budget to when the reports were due to Congress, which allowed OSD reviewers to meet for one hour with representatives of each program office that submitted a SAR.

CAPE O&S analysts report that they rarely participate in post-IOC reviews because of workload constraints, such as the reviews mandated in Section 2441 of the FY 2017 NDAA.

The FY 2016 and FY 2017 NDAAs expanded the scope of analyses traditionally done for MDAP milestone reviews. The analyses described in or implied by the legislation require skills in addition to cost analysis and are typically done during product development as part of the system engineering process and in business-case analyses that address product support requirements. The kinds of analyses described

Table 5.1
Assessment of CAPE Compliance with Law Regarding O&S Cost-Estimating Activities

Cost-Related Requirement for MDAPS	Compliance
CAPE must conduct or approve ICEs and ICAs for all MDAPs and major subprograms in advance of Milestone A, B, or C approval. CAPE must review all cost estimates and cost analyses conducted relating to MDAPs and major subprograms (per 10 U.S.C. 2334).	
Milestone A • An ICE that includes the identification and sensitivity analysis of key cost drivers that may affect life-cycle costs • Planning for sustainment has been addressed • The level of resources required to develop, procure, and sustain the program is sufficient for success (per 10 U.S.C. 2366a) • Goals for procurement unit cost and sustainment cost. (per 10 U.S.C. 2448a)	**PARTIAL** Four CAPE O&S cost analysts conduct ICEs for ACAT 1D programs but cannot also handle ICAs of service-managed programs Traditional estimates have not included analysis of risk drivers, emphasis on planning for sustainment, or estimates of alternative courses of action
Milestone B • Trade-offs among cost, schedule, technical feasibility, and performance objectives to ensure that the program is affordable when considering the per unit cost and the total life-cycle cost • Life-cycle sustainment planning has identified and evaluated relevant sustainment costs throughout the life of the program; and those costs are reasonable and have been accurately estimated (per 10 U.S.C. 2366b) • A life-cycle cost estimate that includes an analysis that identifies and evaluates alternative courses of action that may reduce cost and risk • Goals for procurement unit cost and sustainment cost (Section 2448a)	
Milestone C • A life-cycle cost estimate that includes an analysis that identifies and evaluates alternative courses of action that may reduce cost and risk • Goals for procurement unit cost and sustainment cost (Section 2448a)	
Sustainment reviews Departments conduct sustainment reviews at least every five years after IOC. The reviews assess detailed elements of O&S costs and include an independent life-cycle cost estimate (per 10 U.S.C. 2441).	**NONE** CAPE O&S analysts rarely involved in post-IOC reviews
SARs CAPE must each year review the cost and associated information in SARs (per 10 U.S.C. 2334).	**NONE** Insufficient CAPE O&S personnel

SOURCE: Text of requirements was copied or paraphrased from the indicated sections of 10 U.S.C.

NOTE: Color-coding conveys level of compliance with the law: Yellow denotes partial compliance; red indicates no compliance.

in the FY 2016 and FY 2017 NDAAs are typically done with multidisciplinary teams that include engineers, logisticians, economists, and cost analysts. Some pre–FY 2016 requirements for O&S cost estimating, assessment, and oversight are not being performed due to lack of capacity. Depending on how DoD implements these requirements, CAPE's workload could expand substantially and could require inputs from other organizations in OSD that have primary responsibility for system engineering, personnel, and logistics.

Conclusion

This chapter has highlighted several areas in need of improvement regarding CAPE's compliance with laws concerning O&S cost estimation. There were not enough CAPE O&S cost analysts to handle the workload in the years through FY 2016. This workload did not include the additional cost analysis duties mandated in the FY 2016 and 2017 NDAAs. These laws extend the focus of CAPE's cost analysis to include estimating the cost and adequacy of product-support activities as they relate to O&S costs. Depending on how DoD interprets the requirements, the new duties have the potential to greatly expand CAPE's workload. Chapters Six and Seven provide recommendations to address these issues. Chapter Six focuses on recommendations to address personnel and data issues, while Chapter Seven highlights some broader conclusions from our work and offers recommendations related to OSD's oversight role.

Recommendations for Meeting CAPE's Statutory Responsibilities for O&S Cost Analysis

This chapter provides recommendations designed to help CAPE to fulfill its statutory responsibilities in effect through the FY 2017 NDAA. These recommendations are designed to address personnel and data gaps highlighted in Table 5.1 in Chapter Five.

We stress this caveat to the recommendations: DoD has room for interpretation in implementing its statutory requirements, and this interpretation shapes the resources required. Our recommendations in this chapter for meeting legal responsibilities assume a level of effort and analytical product similar to recent historical experience for a CAPE ICE or ICA.

Personnel

Chapter Three established that the primary driver of cost analysis workload for CAPE cost analysts is conducting ICEs and ICAs for MDAP milestone decisions. Section 825 of the FY 2016 NDAA (Pub. L. 114–92, 2015) amended 10 U.S.C. 2430 to make the component acquisition executive the milestone decision authority for MDAPs except in specific circumstances. Accordingly, we expect that CAPE analysts will conduct fewer cost *estimates* and more cost *assessments* than in previous years. CAPE O&S analysts told us the workloads for estimates and assessments are similar—analysts essentially conduct their own estimates to assess estimates produced by the services. While we expect that the services will likely take the lead in fulfilling the recent requirements for milestone decisions that are beyond those traditional for milestone reviews, CAPE is required to assess ICEs associated with such decisions. These requirements are listed in Table 6.1.

Augmentation of CAPE Staff

Assuming a level of effort and an analytical product similar to what CAPE produces now, we estimate that a total of ten to 16 O&S cost analysts are needed to fulfill all statutory requirements. This range of staffing needs is based on three broad types of tasks. The first broad type of task is the ICEs and ICAs required at milestone reviews.

Table 6.1
Number of CAPE O&S Analysts Needed to Fulfill Legal Requirements

Cost-Related Requirement for MDAPS	Analysts Needed
CAPE must conduct or approve ICEs and ICAs for all MDAPs and major subprograms in advance of Milestone A, B, or C approval.	
CAPE must review all cost estimates and cost analyses conducted relating to MDAPs and major subprograms (per 10 U.S.C. 2334).	
Milestone A • An ICE that includes the identification and sensitivity analysis of key cost drivers that may affect life-cycle costs • Planning for sustainment has been addressed • The level of resources required to develop, procure, and sustain the program is sufficient for success (per 10 U.S.C. 2366a) • Goals for procurement unit cost and sustainment cost (per 10 U.S.C. 2448a)	8
Milestone B • Trade-offs among cost, schedule, technical feasibility, and performance objectives to ensure that the program is affordable when considering the per unit cost and the total life-cycle cost • Life-cycle sustainment planning has identified and evaluated relevant sustainment costs throughout the life of the program, and the costs are reasonable and have been accurately estimated (per 10 U.S.C. 2366b) • A life-cycle cost estimate that includes an analysis that identifies and evaluates alternative courses of action that may reduce cost and risk • Goals for procurement unit cost and sustainment cost (Section 2448a)	
Milestone C • A life-cycle cost estimate that includes an analysis that identifies and evaluates alternative courses of action that may reduce cost and risk • Goals for procurement unit cost and sustainment cost (Section 2448a)	
Sustainment reviews Departments conduct sustainment reviews at least every five years after IOC. The reviews assess detailed elements of O&S costs and include an independent life-cycle cost estimate (per 10 U.S.C. 2441).	2–8
SARs CAPE must review the cost and associated information in SARs each year (per 10 U.S.C. 2334).	0[a]

SOURCE: Text of requirements was copied or paraphrased from the indicated sections of 10 U.S.C.

[a] Assumes a brief review of SARs after the submission of the President's Budget and before the due date to Congress, as currently done by OUSD(AT&L), by a staff of roughly 16 CAPE O&S analysts in addition to their other duties.

The historical workload for these reviews has averaged 16 per year. Each CAPE O&S analyst has been able to accomplish two to three ICEs or ICAs per year. Given the recent expansion in the scope of cost and cost-related tasks done at milestone reviews, we assume each analyst could accomplish two ICEs or ICAs per year. Thus, we estimate eight O&S analysts are needed to support milestone reviews.

The second broad type of task that CAPE O&S analysts should perform is review of cost estimates generated as part of the sustainment reviews conducted by the services. There are roughly 80 MDAPs, and the law requires a sustainment review of each every five years; on average, DoD should thus perform 16 sustainment reviews of MDAPs each year. Although 10 U.S.C. 2334 requires CAPE to review all cost estimates and analyses of MDAPs, there have been too few CAPE O&S analysts to review estimates and analyses associated with sustainment reviews, so there is no historical basis for estimating the potential workload for a typical review. Consequently, the potential range of staff effort required is wide. A cursory CAPE review lasting a month would require roughly one or two additional analysts. A level of effort similar to an ICE or ICA for a milestone review would require eight additional analysts. OSD has latitude in interpreting this requirement, and we are unable to narrow the range of staff effort required.

A third broad type of task is the review of SAR data. Staff in OUSD(AT&L) reviewed SARs (through the December 2017 SAR submission) for consistency and coherency during a one-month period after the President's Budget is submitted and before SARs are submitted to Congress. The review did not include a substantive review of the estimate. The yearly staff equivalent of the AT&L review would be roughly one analyst. Given personnel constraints, we believe that, if staff were added at the upper range of our recommendations to perform the first two kinds of tasks, they could also perform the SAR review for the MDAPs under their purview.

We recognize the mandate to reduce DoD headquarters staff makes sourcing personnel increases difficult. However, given the recent legislative requirement to split OUSD(AT&L) into two distinct entities, additional billets and personnel may be available for transfer to CAPE from former OUSD(AT&L) organizations. Currently, civilian cost analysts in CAPE are staffed at the General Schedule 15 pay grade, which further complicates the ability to source billets and personnel.

We recommend that new staff be hired at lower General Schedule levels and that their skills and experience be developed. Junior and senior analysts would work in teaming arrangements, with junior analysts performing data collection and analysis under the guidance of senior analysts, and senior analysts taking the lead in discussions with outside organizations. This would have two advantages. First, because cost analysis is an occupation for which only the basics can be taught in a classroom and that requires experience to gain mastery, hiring junior staff would enable and encourage more-senior analysts to mentor new analysts. Second, promotion potential within CAPE would facilitate longer tenure and less turnover.

If billets from other OSD organizations are not available, a less ideal solution is that personnel with cost or logistics backgrounds from other organizations be assigned to supplement existing CAPE analysts.

Data

DoD's ability to collect, organize, and analyze cost, logistics, and programmatic data has grown tremendously over the last few decades. The increased capability is the result of persistent leadership and effort throughout the department and advances in computing technology and information systems. Pioneers in the field of cost analysis have recognized the importance of the capability to collect, organize, and analyze data since before the creation of the CAIG:

> A really effective cost analysis capability cannot exist without systematic collection and storage of comparable data on past, current, and near future programs. Even this is not enough. The data must be processed and analyzed with a view to the development of estimating relationships which may be used as a basis for determining the cost impact of future proposals. (Fisher, 1970, p. 77)

Continue Management Support of Existing Efforts to Address Data Gaps

DoD is actively addressing most of the problems with cost and programmatic data we summarized in Chapter Four. For example, it is collecting CSDRs for contractor logistics support costs and is implementing requirements for more-thorough reporting of associated programmatic data in the reports. In our view, one of the most pressing data shortfalls that affects the ability to estimate O&S costs is information on logistics planning and support costs and outputs during system acquisition. These data for legacy systems should be available from the service organizations that sustain the systems, but the data are difficult or impossible for independent cost analysts to obtain. DoD implementation of the provisions in Sections 911 and 912 of the FY 2018 NDAA, which direct OSD to extract data from component business systems and make the data available to CAPE and other DoD offices, could improve this shortfall. CAPE and the services should continue to work to ensure that acquisition contracts require detailed reporting of costs and quantities of unique support equipment, training devices, and initial spares and that data on product-support costs for legacy systems are made available to CAPE.

Make Additional Data Available to Support Product Support and O&S Cost Analyses

We also recommend expanded efforts to obtain additional information, particularly from the services, that would be useful to support O&S cost analyses. Information on the unit cost, quantities, and scope of work of depot maintenance should be routinely provided to CAPE, as one example. Historical data on costs and quantities of unique support equipment, training devices, initial spares, and the adequacy of these levels relative to readiness requirements should be collected from the services on fielded systems.

Obtaining CSDRs and other cost and technical data for the commercial portions of commercial-derivative systems (such as the KC-46 tanker) continues to be a problem for cost analysts and is part of a larger issue concerning government rights to commercial data that remains contentious and unresolved. This problem cannot be solved by DoD alone and would require a change in law.

VAMOSC Systems

OSD and the services are actively addressing issues with service VAMOSC systems. Adequate VAMOSC capability is vital for CAPE and all other DoD cost organizations to perform their duties adequately. Budgeted funding levels as of mid-2017 indicate a decline in funding for VAMOSC systems, and budget reductions targeting headquarters activities and contractor support make it increasingly difficult for the services to maintain these funding levels. Because visibility of DoD O&S costs should be a broad and strategic concern of DoD leadership, we recommend that OSD assume responsibility for funding VAMOSC systems to ensure that adequate funding of service VAMOSC systems is included in the President's Budget request and is fully executed, while the services should continue to manage the systems.

Reliability and Maintainability Data

Key determinants of maintenance costs are the frequency with which equipment needs maintenance (including scheduled and unscheduled maintenance), repair, or replacement (how often) and the associated cost of each event (how much). Although we do not expect the VAMOSC systems to be the primary source of detailed information of this kind, VAMOSC systems provide some insight into frequency of demands and associated maintenance costs for consumable and reparable parts for organically maintained weapon systems. The VAMOSC systems typically provide much less insight into the frequency and associated costs of system-level (e.g., ship, aircraft, or engine) depot maintenance for organically maintained systems. The VAMOSC systems provide little if any insight into how often equipment requires maintenance and how much the maintenance events cost for contractor-supported weapon systems.

Detailed R&M data for weapon systems from their components should be more appropriately archived in a database other than the VAMOSC systems, and such data are available in other databases. An example is the Air Force's Logistics Installations and Mission Support–Enterprise View, which is an information system that consolidates logistics data, including R&M data, from a variety of databases. Such data are useful to the system engineering, test, logistics, and cost communities in DoD. We concur with the recommendation in NRC (2012) that DoD should create a database of performance data from test and operational use, obtained from contractors and the government. Such a database would enable important goals in the development of defense systems, including better system engineering to produce more feasible requirements and improved test design and modeling and simulation. The NRC recognized that such a database could also support analysis to better understand and control maintenance costs because a key driver of maintenance cost is reliability and the associated costs for replacement parts or repair:

> A data archive could support analysis to control and manage a considerable fraction of operations and support costs by revealing and quickly fixing system deficiencies through a failure mode, effects, and criticality analysis, and a failure reporting, analysis, and corrective action system supported by such data collection. (NRC, 2012, p. 54)

Authority

We noted in Chapter Four that the DoDD that describes the duties of CAPE and provides its authority also directs other OSD principal staff assistants and the heads of DoD components to provide CAPE with timely access to any data it requires to perform its duties. Despite this direction, SMEs indicated that CAPE O&S cost analysts sometimes face difficulty in obtaining cost and programmatic data or access to expertise. Because most of the O&S cost workload is driven by estimates prepared for MDAP milestone reviews that are completed within six months, there is little time to resolve problems about getting needed data or expertise. Difficulty in obtaining data is an ongoing problem that is symptomatic of ingrained cultural norms in DoD. The problem was recognized by the House Armed Services Committee in its report on the FY 2018 NDAA:

> The committee is concerned that the Department lags well behind the private sector in effectively incorporating enterprise-wide data analyses into decision making and oversight. The committee therefore believes that a statutory requirement that the Office of the Secretary of Defense, the Joint Staff, and the military departments be given access to business system data is necessary to overcome institutional and cultural barriers to information sharing. (House of Representatives Committee on Armed Services, 2017, p. 169)

The enacted FY 2018 NDAA contains provisions to establish a Chief Management Officer in OSD to manage business systems data in a common enterprise that extracts data from relevant systems in DoD. DoD components are required to provide access to their relevant defense business systems to populate the database (Pub. L. 115-91, 2017). The law explicitly mandates that CAPE have access to these data to perform its duties (10 U.S.C. 2222 (e)(6)(C)). These steps should make it easier for all participants in the acquisition process, including CAPE, to obtain information needed for decisionmaking and oversight.

In a 2012 report, the NRC also acknowledged the need in DoD for enterprise-wide collection of data and its use in decisionmaking and oversight but did not attribute problems in acquiring the data to lack of statutory or regulatory authority. Instead, it concluded that "[m]any of the critical problems in the U.S. Department of Defense acquisition can be attributed to the lack of enforcement of existing directives and procedures rather than to deficiencies in them or the need for new ones" (NRC, 2012, p. 59). We share this view.

Summary

The personnel and data resources that CAPE needs to fulfill its O&S cost analysis duties for MDAPs are small considering the level of personnel resources in OSD and DoD. The main impediment to enabling a more robust role is a belief that OSD should have a minimal and episodic role in the management and oversight of MDAPs. The law is inconsistent in this regard. Provisions in WSARA require CAPE cost analysis at MDAP milestone decisions, and subsequent laws have added to the scope of CAPE's O&S cost analysis duties. Yet recent legislation broke up OUSD(AT&L) and established the services as the default milestone decision authorities for MDAPs.

The laws regarding CAPE's authorities and responsibilities have not been repealed or qualified. Senate Armed Services Committee Chairman John McCain affirmed his expectation of OSD oversight of acquisition in a statement shortly before the FY 2018 NDAA was signed into law (McCain, 2017). Under Secretary of Defense for Acquisition, Technology, and Logistics Ellen Lord has affirmed the need for the newly created Under Secretary of Defense for Acquisition and Sustainment to elevate the importance of sustainment (Lord, 2017).

In Chapter Seven, we make the case for a stronger OSD role in the oversight of product support and O&S.

CHAPTER SEVEN
Recommendations for Improving O&S Outcomes in DoD

This chapter provides the case for a more robust OSD role in overseeing O&S cost and logistics outcomes for DoD systems. It argues for moving from a CAPE role in the estimation of *O&S* costs for MDAPs to ensuring that *product support* efforts during acquisition are executed as planned.

In the first several sections of this chapter, we build the case for this recommendation by explaining the steps in the logic that led to our conclusion:

- O&S cost and logistics outcomes are linked and determined primarily during development.
- Characteristics of the DoD acquisition process inhibit system engineering for better O&S cost and logistics outcomes:
  - Requirements for system engineering for reliability, maintainability, and logistics support for DoD systems have been relaxed.
  - Government system engineering capability has diminished.
  - There are cultural barriers in DoD to cooperation and information sharing.
  - There are inadequate incentives in the acquisition system for improved O&S cost and logistics outcomes.

Taken together, these conclusions support our recommendation for a stronger OSD role in encouraging improved O&S cost and logistics outcomes.

The sections that follow build on findings presented earlier in this report while also highlighting recent support of these ideas from prominent stakeholders that legislate, influence, or implement acquisition policy. In the reports cited here, these stakeholders—in particular, DSB, NRC, and DoD DOT&Es—have provided recommendations for improving the system engineering process in DoD. In this chapter, we cite selected findings and recommendations from these reports that emphasize the link between system engineering efforts during development and O&S cost and logistics outcomes. We reaffirm and advance recommendations focused on improving the ability to estimate O&S costs.

The Need for a More Robust OSD Role in Overseeing O&S Cost and Logistics Outcomes for DoD Systems

O&S Cost and Logistics Outcomes Are Linked and Determined Primarily During Development

The premise that the O&S cost and logistics outcomes of a product or system are linked and determined primarily during development has long been recognized. In this report, we highlight these issues, especially in Chapters One and Two.

This linkage is not unique to DoD or weapon systems. An authoritative textbook on the subject is *Logistics Engineering and Management* by Benjamin S. Blanchard, first published in 1974. The sixth edition describes a systems approach toward logistics that treats the development of a product holistically, including its integrated logistics support, rather than considering each component of the total product separately. A systems approach to logistics requires considering reliability, maintainability, and supportability during the design process. Blanchard's text recognizes that most system O&S costs are determined by decisions early in the design process, which makes it imperative that logistics be considered in the early stages of design to control life-cycle costs. The book provides detailed instruction on how this should be done (Blanchard, 2004).

While the need for system engineering and planning for logistics support is not unique to defense systems, DoD faces a difficult challenge in developing and fielding weapon systems that keep pace with the threat and are reliable, available, and affordable. As we have emphasized throughout this report, the need for increased capability has led to weapon systems that are increasingly complex and generally have multiple hardware and software subsystems or components (NRC, 2015).

The increased capability and complexity of weapon systems has been accompanied by a decline in reliability and availability and a corresponding increase in O&S costs of fielded systems. As summarized in Chapter One, multiple directors of OSD DOT&E have documented the decline in reliability of tested systems since the 1980s. Although public information on the availability rates of weapon systems is scarce, recent DOT&E reports document availability rates lower than goals for recently tested systems, such as LCS and F-35. Both programs are immature, and their availability will likely improve as they mature. Our examination of weapon system availability rates for these and several other MDAPs with large fleets reveals availability rates much lower than were planned during development and lower than current established goals for the systems.

Stakeholders, including multiple DOT&Es, have recognized the link between O&S cost and logistics outcomes and system design. The FY 2013 DOT&E report observed that reliable systems have lower O&S costs because they require less maintenance and fewer spare parts and are more likely to be available to perform their missions (DOT&E, 2014, p. vi). DSB likewise found that a system engineering program that includes reliability, availability, and maintainability during design and development is the most important element in improving the suitability of defense systems (DSB, 2008).

Under Secretary Lord has acknowledged that R&M must be considered early during acquisition design and development (Lord, 2017). Similarly, the House Armed Services Committee report accompanying its FY 2018 NDAA bill acknowledged:

> The design of a major weapon system directly affects its life-cycle sustainment activities and consequently drives its O&S costs. Elements of sustainment that are highly dependent on the system design, namely R&M, are easier and less costly to address during the development of an MDAP than after a weapon system is fielded. Therefore, the committee believes the Department should emphasize R&M in early engineering decisions. (House of Representatives Committee on Armed Services, 2017)

The FY 2018 NDAA codified the requirement for program managers to address R&M in acquisition contracts.

As noted in Chapter One and discussed in greater detail in Appendix A, we measured the growth in estimates of direct unit O&S cost reported in SARs for 20 MDAPs with large O&S costs. In several cases, we saw that, when R&M metrics specified in development were not achieved, the numbers of maintenance personnel assigned to systems and their cost increased. Similarly, maintenance costs for elements such as spare parts and depot maintenance increased. Many of these programs had much lower availability than originally planned. The outcomes appear to be linked.

For MDAPs with higher-than-average growth in unit O&S costs, foundational planning documents that inform estimates often contained

- estimates of unit-level personnel requirements lower than ultimately assigned
- R&M metrics lower than ultimately achieved
- system-level R&M metrics of little use in estimating costs of components or subsystems
- inadequate determination of maintenance requirements.

For programs with higher than average growth in unit O&S cost, the outcomes of the cost-estimating approaches underscore

- the usefulness of an antecedent as a sanity check
- the danger of relying on assumptions specified in foundational documents without critically evaluating the validity of the assumptions
- the value of conducting sensitivity analyses around foundational assumptions regarding personnel requirements and R&M to produce a range of potential O&S cost estimates

- the importance of consistent DoD guidance regarding the elements of cost to include in the estimate.

Characteristics of the DoD Acquisition Process Inhibit System Engineering for Better O&S Cost and Logistics Outcomes

The importance of system engineering and logistics planning to achieve good outcomes in the O&S phase is widely understood. What impediments to these practices to achieve more successful and consistent O&S outcomes across DoD programs exist? Experts suggest several impediments, as described in the subsections that follow:

- Requirements for system engineering for reliability, maintainability, and logistics support were relaxed in DoD.
- Government systems engineering capability has diminished.
- There are cultural barriers in DoD to cooperation and information sharing.
- There are inadequate incentives in the acquisition system for improved O&S cost and logistics outcomes.

Requirements for System Engineering for Reliability, Maintainability, and Logistics Support Were Relaxed in DoD

Chapter Two summarized changes in law since passage of WSARA that addressed O&S and product support for MDAPs. Experts have argued that DoD relaxed its requirements and diminished its capabilities regarding product support beginning in the 1990s. Legislation after WSARA can be understood as a reaction to these changes.

For example, DSB (2008, p. 23) found that, in the 15 years preceding its report, reliability growth methodologies were seldom used in development; military specifications, standards, and guidance were not used; and less emphasis was placed on reliability, availability, and maintainability criteria.

Similarly, the NRC found a shift in DoD policy in the mid-1990s away from designing for high initial quality and a need for better design practices (2015, p. 63). A clear example of this is the cancellation of MIL-STD-1388, *Logistic Support Analysis*, in 1996–1997.

MIL-STD-1388 provided guidance and codified logistics support analysis for DoD. It was established in 1973 and downgraded to a best practice in 1996–1997. When that standard was canceled, DoD adopted commercial standards. The purpose of MIL-STD-1388 was to establish supportability requirements as a key factor in system design and requirements. By accounting for operational support requirements during acquisition development and by preplanning integrated logistics support, DoD intended to reduce life-cycle cost and increase efficiency. MIL-STD-1388 provided a uniform approach that defined support requirements related to system design, support system design, and logistics requirements. It also defined operational phase support (MIL-STD-1388-1A, 1993).

Government Systems Engineering Capability Has Diminished

SMEs have observed a decline in government expertise in system engineering since the mid-1990s, when acquisition reform policies were implemented and since DoD staffing levels decreased (NRC, 2015, p. 22). Quantifying the decline in personnel with expertise in system engineering and R&M is difficult because there is no occupational series in the government for such personnel that can be tracked over time. There are occupational series for logistics management and various engineering specialties, although not for system engineers. in addition, some of this expertise has been provided by contractor personnel, the numbers and specialties of which cannot be tracked over time.

Gross measures of the scale of the reductions in the acquisition workforce are indicated in legislation and in a DoD Inspector General audit. For example, Section 906 of the FY 1996 NDAA required the Secretary of Defense to plan a 25-percent reduction in the acquisition workforce over five years (Pub. L. 104–106, 1996). Subsequent NDAAs specified numbers of personnel to be reduced each year. In 2000, the DoD Inspector General found the following:

> Using the congressional definition of the DoD acquisition workforce, DoD reduced its acquisition workforce by about 50 percent from the end of FY 1990 to the end of FY 1999; however, the workload has not decreased proportionately. There is cause for serious concerns related to mismatches between the capacity of the reduced workforce and its workload; adverse performance trends; implications of skills imbalance and projected high attrition; and disconnects in workforce planning. (DoD, Office of the Inspector General, 2000, p. 4)

DoD has taken steps to rebuild its acquisition workforce since the drawdowns noted earlier. But it takes time to acquire and train personnel and for them to acquire experience with complex programs. A 25-percent reduction in headquarters staffs from FY 2016 levels, mandated in Section 346 of the FY 2016 NDAA (Pub. L. 114–92, 2015), will cause further reductions in acquisition staff in OSD and the services. In 2017, there were indications of the need to improve skills in planning for product support.

We found indications of logistics planning problems that the services had discovered during integrated logistics assessments of their programs and in our own observation of programs. A Naval Air Systems Command analysis of integrated logistics assessments of 17 programs found systemic deficiencies in maintenance planning (including depot planning), technical data, life-cycle sustainment plans, integrated master schedules, and program funding levels. The analysis found that program personnel may know how much funding they *have* for logistics but do not know how much funding they *need*. CAPE cost analysts have the same problem in assessing the adequacy of logistics support resources for historical programs and estimating costs of logistics support for prospective programs. Naval Air Systems Command is taking action to address the findings and improve knowledge among its logisticians by training, sharing expertise from centralized departments to program offices, and developing standardized tools and templates (Naval Air Systems Command, 2017).

Our observation of Army independent logistics assessment briefings revealed deficiencies in logistics planning that resulted in maintainability, reliability, and funding issues for fielded systems.

More broadly, we reviewed assessments of the logistics status of MDAPs written by analysts in the Office of the Assistant Secretary of Defense for Logistics and Materiel Readiness. The assessments indicated logistics planning problems across the services and commodity groups (ships, aircraft, combat vehicles, etc.) that resulted in problems, including inadequate funding for spare parts, lack of depot repair capability, and other issues that affected system availability. We incorporated these assessments into our review of MDAPs with large increases in unit O&S costs, as summarized in Appendix A.

There Are Cultural Barriers in DoD to Cooperation and Information Sharing

The drawdown in the acquisition workforce, coupled with the increasing complexity of new DoD programs, makes it difficult for program offices to staff their programs with experienced and highly skilled personnel in system engineering and logistics. Current indications of problems in logistics planning reflect this. A potential way to address this problem is to share expertise and input across different organizations and functional areas.

NRC held a workshop that identified the importance of obtaining SME input on the maturity of technology used in new programs. Program managers in industry avoid including unproven technology in new programs. While DoD has managed technology maturation successfully in some programs, the Council wrote that

> What is needed is a way to instill a willingness to acquire independent expert input and a collaborative spirit in those leading future programs. Such a culture is the responsibility of the most senior DOD acquisition executives and of the secretary of defense. The problems result from the different cultures and practices of the different participants in the requirements development process, the acquisition process, and the resource allocation process—not in stated DOD policies and procedures contained in DOD directives. (NRC, 2012, p. 36)

Our recommendation at the end of this chapter provides a way to acquire independent expert input for a limited number of MDAPs.

There Are Inadequate Incentives in the Acquisition System for Improved O&S Cost and Logistics Outcomes

Compounding the problems described above is the competition to approve and fund new acquisition programs, coupled with a lack of adequate incentives to improve O&S cost and logistics outcomes. Resources are always limited, and programs are therefore essentially in competition with each other for funding.

Program managers that are strapped for cash during the acquisition phase may lack the funding to do planning for product support. These pressures may cause the government program managers to prioritize acquisition costs and proceeding through acquisition milestones over sustainment costs. Although we do not claim these problems afflict most DoD programs, these characteristics of the acquisition system are widely known and have long been recognized. We cite recent descriptions of the problem from prominent sources.

Section 809 of the FY 2016 NDAA mandated the creation of an advisory panel to look at the defense acquisition process. The panel submitted its interim recommendations to Congress in May 2017. The panel found the following:

> Contractors sometimes use unrealistically low cost estimates to win contracts; program representatives use low estimates to argue for approval of the system against competing systems. Such optimism in cost, schedule, and performance often leads to cost overruns, schedule slips, and capability gaps or shortfalls. Incentives are needed that promote more candor in presenting programs to Congress and senior leaders in DoD. (Section 809 Panel, 2017)

As noted in Chapter Two, House Armed Services Committee Chairman Mac Thornberry has recognized that the acquisition process provides incentives for near-term trade-offs of cost, schedule, and performance that reduce costs in acquisition at the expense of higher O&S costs (Thornberry, 2017).

GAO has similarly observed

> that there are strong incentives within the culture of weapon system acquisition to overpromise a prospective weapon's performance while understating its likely cost and schedule demands. Competition with other programs vying for defense dollars puts pressure on program sponsors to project unprecedented levels of performance (often by counting on unproven technologies) while promising low cost and short schedules. These incentives, coupled with a marketplace that is characterized by a single buyer (DOD), low volume, and limited number of major sources, create a culture in weapon system acquisition that encourages undue optimism about program risks and costs. (GAO, 2016, p. 27)

In their defense classic, *How Much is Enough: Shaping the Defense Program 1961–1969*, Enthoven and Smith observed the same tendencies in DoD in the 1960s and argued for the role of independent analysts, such as those in CAPE, as a check (Enthoven and Smith, 2005).

While there are powerful incentives for those in acquisition organizations to underestimate program costs and shortchange near-term efforts that would reduce sustainment costs, there are no positive incentives for better program management as there are in commercial organizations. GAO (2015) and NRC (2012) contrast the role and incentives of program managers in commercial organizations against those in DoD:

> The commercial model, in which good program outcomes can be achieved with a more streamlined oversight process, includes a natural incentive that engenders efficient business practices. Market imperatives incentivize commercial stakeholders to keep a program on track to meet business goals. In addition, awards and incentives for managers are often tied to the company's overall financial success. As a result, commercial managers are incentivized to raise issues early and seek help if needed. They know if the program fails, everyone involved fails because market opportunity is missed and business revenues will be impacted. Commercial product development cycle times are relatively short (less than 5 years), making it easier to minimize management turnover and to maintain accountability. DOD's acquisitions occur in a different environment in which cycle times are long (10 to 15 years), management turnover is frequent, accountability is elusive, and cost and schedules are not constrained by market forces. (GAO, 2015, p. 28)

An earlier GAO report observed that the average tenure of a DoD program manager is 17 months, while programs can have life cycles exceeding 20 years. The report also noted that no single person or organization in DoD controls sustainment costs, further inhibiting the ability to hold anyone accountable (GAO, 2010).

The NRC contrasted the role of a DoD program manager with a counterpart in industry in the same way as GAO. The NRC emphasized that the tenure of a program manager in industry spans the entire process from product planning and design to fielding and product support. The long tenure enables seamless transition through the stages of product development and allows the program manager to be held accountable for outcomes:

> In contrast, in DOD the tenure of a program manager rarely covers more than one phase of a project, and there is little accountability. Moreover, there is little incentive for a DOD program manager to take a comprehensive approach to seek and discover system defects or design flaws. (NRC, 2012, p. 9)

The provisions in Section 834 of the FY 2018 NDAA require program managers to specify R&M metrics in acquisition contracts and partially address these issues. However, achieving logistics outcomes also requires adequate product support in such elements as peculiar support equipment, peculiar training equipment, publications and technical data, and initial spares. Adequate support will require accurately estimating the cost of these elements and funding them.

It is important to realize that the incentives of acquiring organizations and program managers are inherent to the DoD acquisition system and are barriers to improved O&S outcomes. Therefore, proposed solutions that would give more responsibility or authority to these organizations cannot fully address the problem.

Recommendation: OSD Should Strengthen the Role of CAPE to Encourage Improved O&S Cost and Logistics Outcomes

Expectations for DoD's ability to manage and control system O&S costs should be tempered by DoD's past efforts and informed by the barriers to success knowledgeable stakeholders have identified. The department has implemented strategies during the acquisition process to control weapon system costs over the past several decades. The DTC and Cost as an Independent Variable policies, which date back to the early 1970s, were largely ineffective (Kneece et al., 2014). Its early SARs described the F-35 program, which was one of the Cost as an Independent Variable pilot programs, this way:

> The program was structured from the beginning to be a model of acquisition reform, with an emphasis on jointness, technology maturation and concept demonstrations, and early cost and performance trades integral to the weapon system requirements definition process. (DoD, 2001)

The estimated unit O&S cost of the F-35A has more than doubled since development began.

Kneece and colleagues attributed the failure of previous policies to control costs to a few probable causes:

- lack of emphasis on early sustainment concept formulation and rigorous system engineering that considers these needs
- uncertainty in O&S cost estimates, especially early in the acquisition process
- difficulty motivating acquisition participants to control O&S costs
- lack of high-level and persistent management attention and processes for reassessing affordability (Kneece et al., 2014).

Stronger OSD Role in Management of Sustainment

We recommend a stronger OSD role in management of sustainment to address these problems and to improve cost and logistics outcomes in the O&S phase. A key feature of our recommendation is continuous involvement with MDAPs by CAPE cost analysts and OSD personnel from systems engineering and logistics offices, rather than CAPE's current brief and episodic involvement preceding milestone reviews. We realize there are limited personnel in OSD and pressure to further reduce staff, and we also realize many stakeholders in Congress and DoD prefer more service responsibility for acquisition and less OSD involvement. We therefore propose a stronger OSD role for selected programs on a trial basis. The programs might be selected according to such criteria as estimated O&S cost or by membership in a portfolio of programs of a similar commodity, such as aircraft that could benefit from the cross-service expertise that OSD could facilitate. The trial could be evaluated initially by OSD and service participants by assessing sustainment issues raised and resolved under the trial process as opposed to the more traditional oversight process. It could take many years before actual data are available for an assessment based on cost-estimating and product-support outcomes.

OSD Subject-Matter Experts Available Continuously to the Program

A second key feature of our recommendation is to provide substantive input to CAPE cost analysts in such areas as personnel requirements for support of weapon systems; system and component R&M; requirements for logistics support, including facilities and depot repair capabilities; and contracting. Expertise in contracting is required to ensure that contracts appropriately reflect requirements for product support activities and outcomes. The best way to do this would be to augment CAPE staff with experts in these areas because CAPE's charter for independence fosters objectivity in providing substantive inputs and assessments.

Given the pressure to reduce OSD staffing, a less desirable solution would be to establish a process and issue a formal DoD directive or instruction to require OSD SMEs from offices other than CAPE to provide input to CAPE cost analysts for their assessments of MDAPs at milestone reviews. A model for what we envision is the Naval Air Systems Command's Estimating Technical Assurance Board process. The purpose of the process is to ensure the provision of credible technical inputs for the command's cost estimates. The process requires that general officer or senior executive service–level leaders in technical competencies validate the technical inputs used in cost estimates at major reviews (Naval Air Systems Command Instruction 5223.2, 2010).

Personnel Resources Needed to Support This Recommendation

A more robust OSD role would require additional cost analysts in CAPE, as recommended in Chapter Six. It could also require more personnel in OSD in system engineering, including planning for logistics support. DSB and NRC have recommended specific actions in this area. As one promising example, SMEs in the Office of the Deputy Assistant Secretary of Defense for Systems Engineering described advances in modeling for reliability, availability, maintainability, and cost that can improve the analysis of these relationships before and during system development. RAND has not assessed the validity of the tools. The tools are not used uniformly across DoD, and the department may not have an adequately skilled workforce to apply the them. A more robust OSD role could help transfer this knowledge between and among the services and encourage broader stakeholder involvement in the development of program cost estimates. These changes would address the first impediment to improved outcomes of inadequate emphasis on up-front system engineering efforts to incorporate product support concerns.

Expected Outcomes from Implementing This Recommendation
Continuous CAPE involvement coupled with an organizational mandate to strengthen technical inputs to cost estimates should reduce the impediment of large uncertainty in O&S estimates.

Continuous CAPE involvement could strengthen the incentives of program managers and others in the acquisition organization to address sustainment in two ways. First, budget constraints in development often lead program managers to fund near-term development at the expense of planning for logistics support. Cost estimates that are more well-informed of requirements for logistics support could help ensure that adequate funds are budgeted for these efforts during development. Second, continuous involvement of OSD staff would allow participation in program reviews with the weapon system contractor to ensure R&M engineering and logistics planning efforts are being executed. Episodic involvement prior to milestone reviews is too infrequent.

More robust and well-informed input from CAPE and other OSD offices at milestone and other reviews should give OSD leadership greater confidence and capability to provide the high-level and persistent management attention to these areas that has been lacking in the past.

A Stronger OSD Role in Sustainment Oversight Need Not Lengthen Cycle Times
Defense stakeholders in Congress and DoD have a strong interest in shortening acquisition cycle times to develop and field weapon systems. Cycle times measured from Milestone B to IOC (or from Milestone C for programs using mature technology and without a Milestone B) have increased over the last few decades to a current mean average of seven years for MDAPs. The increases are probably caused by increases in the complexity and capabilities of the programs (OUSD[AT&L], 2016, p. xxxvii). A natural concern with increased attention to sustainability during acquisition is the potential to increase cycle times. Some perspective on this concern is offered by consideration of some extreme cases of long cycle times highlighted in the OUSD(AT&L) 2016 Annual Report.

The six MDAPs with the longest cycle times were those for the F-35, Advanced Extremely High Frequency Satellite, MQ-8 Fire Scout, F-22, Excalibur precision 155-mm projectile, and Advanced Threat Infrared Countermeasures/Common Missile Warning System, which have cycle times from 14 to 15 years. The six programs with the longest cycle times all began at Milestone B and involved lengthy product development and testing requirements. (OUSD[AT&L], 2016, p. 46). Our review of literature on the six programs with the longest cycle times did not identify consideration of sustainment issues as a contributor to the extended program durations. The CH-53K helicopter program is similarly lengthy and is currently in development, with an IOC scheduled for 2019 and cycle time of 14 years. GAO's reviews of the program identified reasons for the lengthy development schedule that did not include inordinate planning for sustainment. In fact, one of the GAO reports noted that two of the program's R&M metrics have been relaxed (GAO, 2011).

For some development programs for which there is an urgent operational need, proper planning for logistics support could slow fielding. Drew and colleagues described this dilemma with Air Force accelerated fielding of unmanned aerial vehicles and acknowledge that decisionmakers must balance the urgent need for a capability and the effects of rapid development on the long-term support of that capability (Drew et al., 2005). However, for most MDAPs, including the seven with extraordinarily long cycle times just cited, we found no evidence that inordinate attention to product support has delayed fielding.

Our recommendation for a stronger OSD role need not increase requirements for documentation for MDAPs or lengthen cycle times. Rather, continuous involvement is intended to facilitate rapid sharing of information on an informal basis.

Summary

The need to keep pace with the increasing capabilities of potential adversaries has led to the increasing complexity and capability of DoD systems. Achieving desired cost, R&M, and availability outcomes for these new systems is difficult under the best of circumstances. Legislation enacted in NDAAs since 2009 indicates congressional interest in improving the outcomes and has mandated analyses and management activities that various DoD organizations must perform, with the intent of improving logistics outcomes and O&S costs. A powerful and recurring theme in the legislation is a realization that product support activities early in acquisition largely determine the O&S costs and logistics outcomes of fielded systems. The legislation requires the independent estimation of these costs and their adequate funding. We believe the recent legislation since 2009 regarding independent cost estimating is an important shift in emphasis from a more traditional focus on O&S costs to broader product support activities and costs that span the acquisition life cycle.

An equally evident theme is the preference for the military services to manage the acquisition of MDAPs. Although legislation stipulates that the default milestone decision authorities for MDAPs are the service acquisition executives and that additional product support activities and analyses are the responsibilities of the military services, 10 U.S.C. 2334 still requires CAPE to conduct or approve ICEs and cost analyses for all MDAPs and major subprograms in advance of Milestone A, B, or C approval.

Paradoxically, the new emphasis on CAPE's estimation of product support activities and costs and the attendant expansion in scope of its responsibilities have not been accompanied by additional personnel to augment the four O&S cost analysts. Independent assessments of the adequacy and cost of product support activities require input from independent substantive experts. Cost analysts alone cannot be expected to possess this expertise. In short, legislation since WSARA has expanded CAPE analysis and estimation of O&S activities but has also moved resources and responsibilities away from OSD.

OSD can implement the recommendations in this chapter within its current legal authority and with minimal resources. The reaction from the services and MDAP program managers is difficult to predict, but the prospect for shared cross-service knowledge and collaboration in areas critical to program success should be attractive.

The recommendations in this chapter allow DoD to fulfill its responsibilities for independent estimation or approval of cost estimates for MDAPs at milestone decisions. For selected MDAPs for which OSD oversight would be especially beneficial, the recommendations would enhance product support as intended in recent legislation.

O&S Cost Estimates in Selected Acquisition Reports

In this appendix, we present the results of our analysis of DoD unit O&S cost-estimating results as reported in SARs. Unit O&S costs are the O&S costs defined in the standard DoD cost element structure to operate one unit (e.g., one ship, ground vehicle, or aircraft) for a defined time (generally one year). Our use of the estimates in SARs raises some issues and requires explanation to understand and appreciate the results that follow.

As described in the following section, SARs contain DoD's official cost estimates for MDAPs and, by law, are transmitted to Congress. Congress has used SARs as an important source of information on the cost, schedule, and technical performance of MDAPs for over 40 years. For this reason, estimates in SARs are a logical choice for assessing DoD cost-estimating results over time.

Estimates in SARs are prepared by the program office for each MDAP. The only exception to this is the F-35 program, for which the estimates of O&S costs in its SAR have been prepared by CAPE in recent years. For this reason, estimates in SARs are poor indicators of CAPE cost-estimating results and should instead be viewed as indicators of the historical accuracy of component cost estimates.

This appendix includes analyses of how the unit O&S cost estimates have changed over time. For many of the programs, the measurement of growth began with estimates made in the 1990s. Many things affecting O&S cost estimates have changed since then, including guidance on what elements of cost should be included in O&S estimates, the level of OSD oversight of SARs, and the ability of VAMOSC systems to capture comprehensive O&S costs for legacy systems to inform estimates. Thus, the findings on changes in O&S cost estimates must be tempered by the realization that these results imply little or nothing about current cost-estimating capabilities.

In the last section of this appendix, we examine a variety of sources of information to seek to understand why unit O&S cost estimates for some MDAPs increased more than average. An appreciation of how much O&S estimates have changed, combined with an understanding of why they changed, offers lessons about the cost estimation and management of sustainment in DoD.

Introduction to Selected Acquisition Reports

DoD established SAR reporting in the 1960s. The first reports contained estimates of cost, schedule, and performance for selected defense acquisition programs. Cost estimates included development, procurement, and military construction costs. By 1970, GAO was using the information in SARs to convey information on the status of defense acquisition programs to Congress (GAO, 1970). The Armed Forces Appropriation Authorization, 1972, (Pub. L. 92-156, 1971) required DoD to begin submitting annual reports on weapon system development and procurement schedules and costs and the results of operational testing. In 1975, the Department of Defense Appropriation Authorization Act for FY 1976 required DoD to submit the reports on a quarterly basis, specified development and procurement dollar thresholds that defined programs as major programs, and used the term *selected acquisition reports* (Pub. L. 94-106, 1975). In 1985, the Department of Defense Authorization Act added the requirement for SARs to include a full life-cycle cost analysis (Pub. L. 99-145, 1985), meaning that O&S costs would be included. The law specifying the content of SARs and the requirement for DoD to provide them to Congress is in 10 U.S.C. 2432, Selected Acquisition Reports.

SARs provide a long historical record of acquisition and O&S cost estimates for MDAPs.[1] SARs provide estimated procurement costs and quantities, including the estimated average procurement unit cost for the MDAP. O&S costs are presented for the entire life cycle. More useful for our purposes, the SARs also provide an estimate of unit O&S cost.

The unit cost estimate is provided in DoD's standard cost element structure for O&S costs. This cost element structure and some definitions have changed over the years, complicating analysis of O&S costs over time. The most recent cost element structure has been in effect since 2014. The six major O&S cost elements are

1.0 Unit-Level Manpower
2.0 Unit Operations
3.0 Maintenance
4.0 Sustaining Support
5.0 Continuing System Improvements (hardware and software modifications)
6.0 Indirect Support.[2]

---

[1] OUSD(AT&L) has established a repository of SARs in the Defense Acquisition Management Information Retrieval (DAMIR) system. The database contains reports dating to the mid-1990s. The system is not available to the general public.

[2] This element may be excluded from the SAR unit cost because of issues in obtaining this data for legacy systems.

The structure contains additional levels of detail in a hierarchical fashion below each of these six elements, although the lower levels of detail are not shown in SARs.[3]

Although SARs have long included estimates of O&S costs, the estimates were not necessarily updated regularly. In December 2008, OSD issued a memorandum to the military services directing that they update and report sustainment metrics on a quarterly basis, including costs for MDAPs. The memorandum provided the procedures for doing so, which involve electronic updates and data sharing through DAMIR (OUSD[AT&L], 2008). Although the memorandum does not mention SARs, SARs are updated and retained in DAMIR.

Criticism of DoD O&S Cost Estimates in Selected Acquisition Reports

This section summarizes criticisms in literature and other published reports of O&S cost estimates in SARs. We revisit these criticisms at the end of the chapter, after presenting the findings from our analysis of the estimates.

A 2012 GAO report assessed the O&S cost estimates in SARs from 2005 to 2010. GAO found several inconsistencies among the SARs:

- the source of the cost estimate was often unclear
- the units of measurement varied (such as cost per aircraft, per squadron, or per fleet) to portray average unit cost
- the explanatory narrative, such as tracking changes over time and identifying cost drivers, was often lacking
- the frequency of SAR updates varied.

In addition, GAO found almost one-half of the estimates for the SARs they sampled to be unreliable. For example, programs omitted costs that should have been included, and one program reported current and projected funding, rather than estimated O&S cost requirements. GAO provided recommendations for clearer guidance and more thorough review of SARs before they are submitted to Congress (GAO, 2012).

A different group of researchers (Ryan et al., 2012) found similar and other problems in developing a methodology to assess the accuracy of O&S cost estimates in SARs:

- SAR cost categories do not map cleanly to specific inflation categories, so it is not possible to compare SAR estimates that are provided in different base-year dollars, even for the same program.

---

[3] The cost element structure is provided and described in CAPE's *Operating and Support Cost Estimating Guide,* (OSD CAPE, 2014).

- The Army's VAMOSC system does not contain all the elements of O&S costs reported in SARs and therefore cannot be used to compare actual O&S costs with estimates reported in SARs.
- Most programs had inconsistencies between values and the text, unstated assumptions, incorrect units of measurement, and other problems.

Notwithstanding the problems with SAR reporting and comparing SAR estimates to costs reported in VAMOSC systems, researchers have examined the O&S estimates in SARs. Jones et al. (2014) examined the ratio of total O&S costs to total acquisition costs. The authors found that the average O&S cost as a percentage of life-cycle cost was 50 to 55 percent, with significant deviations from the average.

Ryan and colleagues examined the accuracy of unit O&S cost estimates in SARs of 37 Air Force and Navy MDAPs. The most recent SAR data in their sample were from 2010. The authors compared estimates of unit O&S costs in SARs to actual unit O&S costs reported in AFTOC and Navy VAMOSC. They found that the mean errors in average unit O&S cost estimates were 40 to 50 percent and that the SAR estimates did not improve with time (Ryan et al., 2013).

RAND Analyses Performed on SAR Cost Data

This section contains the results of five analyses of cost estimates reported in SARs. We conducted these analyses to address five questions:

- How much of the life-cycle cost of MDAPs consists of O&S costs?
- Do unit O&S estimates in SARs become closer to actual costs over time?
- How much have estimates of unit O&S costs changed over time when normalized for usage?
- Is there a difference in the degree of change in the estimates of unit O&S costs in SARs according to the complexity of the weapon system?
- Has the frequency of change in SAR O&S estimates changed since WSARA?

We obtained SAR data from the DAMIR system for all MDAPs that submitted SARs in December 2015 and updated the data for these programs when the December 2016 SARs became available. To this list, we added two Air Force programs for which our sponsor asked us to conduct case studies, the F-22 fighter aircraft and the RQ-4 Global Hawk remotely piloted aircraft. These two programs had stopped SAR reporting in 2010 and 2014, respectively, so were not in the initial list of programs we selected. We removed programs for which no O&S costs had been estimated, leaving 82 programs, with 16 Army programs, 36 Navy programs, 28 Air Force programs, and two DoD programs. Assumed service lives varied from 10 to 50 years, with an average of 26 years and an O&S dollar-weighted-average of 32 years.[4] Programs report costs in various base-year dollars. For some of the analyses we conducted it was necessary to convert the various base-year estimates to a single constant-year dollar basis. We converted the costs using procurement and Operations and Maintenance (O&M) deflators from the OSD *National Defense Budget Estimates for FY 2017* (OSD[Comptroller], 2016). We acknowledge the imprecision in normalizations of SAR data to a constant-dollar basis, as pointed out in literature. Nevertheless, one must bear in mind that the estimates embody expectations by estimators in various program offices of price changes decades into the future, so the entire endeavor is inherently imprecise. Analytic results must be understood in this context.

Several caveats apply to the SAR data and our findings:

- Normalization of SAR costs to the same constant-dollar basis is imprecise.
- The guidance on what SARs should contain has changed over time, and interpretation of the guidance has differed over time and among programs. This affects the O&S costs estimated.
- DoD's portfolio of MDAPs changes frequently. Findings reported in this appendix reflect a snapshot in time.
- Our tracks of changes in unit O&S cost estimates include a subset of MDAPs and are not representative of all MDAPs. The tracks include adjustments for usage and known changes in programs that would affect O&S costs. Tracks of unadjusted data would provide slightly different results.

O&S Proportions of Life-Cycle Cost

The first analysis was a simple calculation of the proportions of estimated lifetime acquisition and O&S costs by type of system. The intent of these calculations is to indicate the relative importance of O&S costs in the MDAP life cycle. An additional insight is how the proportions vary according to the commodity type of the system (aircraft, ship, missile, etc.).

For this analysis, we calculated the proportions of acquisition and O&S costs based on their estimated lifetime costs in constant FY 2017 dollars. Thus, the results can be thought of as dollar-weighted, with the costliest programs affecting the DoD total more than the less costly programs. Results are shown in Table A.1.

The left-hand column of Table A.1 indicates the type of commodity. Most types are self-explanatory. "Other" includes naval aviation and shipboard systems. "O&S Proportion" indicates the percentage of each system's life-cycle cost that is for O&S, with the remainder consisting of acquisition costs. The right-hand column indicates

---

[4] The dollar-weighted average was calculated as the sum of the product of the service life of each program and its estimated lifetime O&S cost, divided by the lifetime O&S cost of the programs. We calculated it to show that the programs with the highest lifetime O&S costs tend to have longer service lives than average.

Table A.1
O&S Proportion of MDAP Life-Cycle Cost by Commodity Type

MDAP Commodity Type	Count of MDAPs	O&S Percentage of Life-Cycle Cost	Percentage of Total DoD MDAP O&S
Other	2	73	1
Ground combat	3	68	4
Helicopter	9	68	12
Aircraft	17	64	54
Ship	8	62	20
Radar	3	46	0
Command, control, communications, and intelligence	12	42	2
Missile	10	36	3
Submarine	1	34	3
Satellite	9	21	1
Munitions	6	17	0
Avionics	1	17	0
Booster	1	2	0
Total	82	59	100
Air, land, and sea-based platforms	39	63	93

SOURCE: RAND analysis of data from DAMIR.

the percentage that each commodity type's O&S is of the total O&S cost of all 82 MDAPs in this calculation.

The O&S proportions by type of commodity are largely intuitive. The O&S proportions are higher for aircraft, ships, and ground-based vehicles than for other types of systems. The average proportion of O&S costs for these platforms is 63 percent of life-cycle costs. The exception to this intuition is the low proportion of O&S costs for the submarine system, which is nuclear-powered. Part of the reason for the relatively low proportion of O&S costs for the submarine is that nuclear power refueling costs are not reported in the SAR. For all 82 programs in this sample, O&S costs accounted for 59 percent of their life-cycle cost.

Changes in O&S Unit Cost Estimates for MDAPs by Program and Characterized by System Complexity

We assessed trends in estimates of MDAPs reported in SARs to determine whether the changes in estimates were significantly different according to their complexity.

We began by tracking the estimates of the 26 MDAPs with the largest life-cycle O&S costs from the total of 82 MDAPs. The selected programs are not representative of all MDAPs. The selected programs are predominately ground vehicle, aircraft, and ship programs because these programs tend to have the most O&S costs.

Of the 26 programs with large O&S costs, we included in this analysis MDAPs that had reported their cost estimates in SARs for at least five years. We tracked the unit O&S estimates for each program from the first SAR available in the DAMIR system to the December 2016 SAR. The exceptions were the F-22 program, which was tracked to its last SAR in 2010, and the Global Hawk program, which was tracked to its last SAR in 2014.

In recognition that some programs change significantly in scope over time in ways that affect cost estimates, we read the descriptions and executive summaries of the programs in the SARs to screen for such changes. On finding significant changes in the scope of the program that would affect costs, we changed the calculations of cost growth to begin with the first year that the revised cost reflecting the change in scope was reported. The AH-64E, CH-47, H-1 Upgrades, and MH-60R programs experienced significant changes in scope, and we adjusted the first year of their cost tracks as indicated.

We also adjusted the O&S cost estimates for programs that had a significant change in the usage rates underlying the estimates because usage rates affect unit O&S costs. For these programs, we assumed the two cost elements of Unit Operations and Maintenance would change in direct proportion to the changes in usage and adjusted these elements accordingly, so that their costs in the last year tracked would reflect the same usage rate as in the original estimate. However, underlying usage rates were not always reported in the SARs, so no adjustment could be made for programs that did not report them.

We characterized the MDAPs as having one of three levels of complexity. Our perspective in characterizing the MDAPs was that of a cost analyst comparing the MDAP to antecedent systems. The lowest level of complexity is programs that are modifications of existing platforms or programs that represent minor changes to existing systems, which we term *modification* programs. The highest level of complexity consists of programs that embody new capabilities or technologies and for which antecedent systems do not provide a solid foundation for cost estimating. We term these *new* programs. Between these two extremes are *mixed* programs, which generally have similar antecedents but also contain enough new technology or capability to present a challenge to cost estimation.

The process of characterization is clearly subjective. We tried to minimize the effect of the subjectivity by having three individuals familiar with defense acquisition and cost estimating characterize the systems independently of each other. We retained programs for which two of the three individuals characterized them the same way. This selection process reduced the list to the 20 programs shown in Table A.2. The 20 pro-

Table A.2
Change in Unit O&S Cost Estimates in SARs

Type	Program	First Year	Change by Program (%)	Change by Type (%)
Mixed	CH-53K	2005	55	45
	DDG 51	1997	25	
	E-2D AHE	2003	29	
	EA-18G	2003	10	
	P-8A	2004	86	
	RQ-4A/B Global Hawk	2001	94	
	SSN 774	1997	17	
Modification	AH-64E Remanufacture	2009	29	49
	C-130J	2003	274	
	CH-47F	2004	7	
	H-1 Upgrades	2008	12	
	HC/MC-130 Recapitalization	2010	13	
	KC-130J	2010	23	
	MH-60R	2003	3	
	PIM	2011	64	
	UH-60M Black Hawk	2001	16	
New	F-22	1997	206	131
	F-35	2001	132	
	LCS	2010	57	
	V-22	1997	129	
	Average of all programs		64	

grams account for 73 percent of the O&S costs reported in SARs for the 82 MDAPs summarized in Table A.1.

Because of the subjectivity in categorizing the programs, we hypothesized that there would be a significant difference in the average growth of the unit O&S estimates between the modification and new programs, but not necessarily between the mixed programs and the other two categories.

The absolute value of the changes in unit O&S estimates averaged 49 percent for modification programs, 45 percent for mixed programs, and 131 percent for new programs. Because there are few programs in each group and wide variance in growth among programs within each group, the group averages are strongly influenced by individual programs. For example, the C-130J is a clear outlier in the group of modification programs, and our track of its estimated unit O&S costs showed that the estimates were far below the actual costs reported in AFTOC and for the actual costs of its antecedent. The average change for the group of modification programs drops to 21 percent when the C-130J is excluded. We conducted t-tests of statistical significance of the difference in the mean changes and found that the mean change of the unit O&S cost of new programs was significantly different from the changes in the unit O&S cost of modification or the mixed programs. The difference in changes between the modification and mixed programs was not significant as determined by a t-test at the 95 percent confidence interval.

Over one-half of the 20 programs listed antecedents and reported the unit O&S costs of the antecedents in the SARs. For these programs, we tracked and measured the change in the unit O&S cost of the antecedents. The antecedent unit O&S costs changed an average of 50 percent from the first reports tracked, or roughly as much growth as for the modification and mixed programs. This finding surprised us. It suggests that much of the growth in O&S cost estimates is attributable to effects common to both existing and new systems. Three drivers of O&S costs affected all programs examined:

- Actual *personnel compensation* and *maintenance* costs for DoD weapon systems in general increased faster than the rate of inflation in the economy or in the OSD Comptroller financial management guidance for budget projections applicable when many of these estimates were prepared.
- *Fuel* costs were much lower through 2004, at less than a dollar a gallon in then-year dollars, than they are now.

Trends of O&S Estimates Toward Actual Costs

A fundamental question is whether cost estimates accurately predict actual costs. Although we would like to be able to address this question for the O&S estimates in SARs, it is impossible to do so authoritatively because SARs report estimates of lifetime O&S costs, which cannot be known with certainty until the end of a program's life. Total lifetime O&S costs are heavily influenced by assumptions about the service life of the program and number of units supported, both of which are highly uncertain. These uncertainties can be mitigated by assessing estimates of average unit O&S costs, rather than total lifetime O&S costs.

Even with estimates of average unit O&S costs, other issues complicate the assessment. One complication is that planned rates of usage can change (for example, flying hours per aircraft per year or ship steaming hours per year), which affect unit O&S costs. The assumptions for usage rates for most systems are based on the requirements for peacetime training, which are known with fairly high confidence. However, actual usage rates can change dramatically from planned peacetime rates for systems used in contingency operations. Remotely piloted vehicles are a good example of this kind of unpredictability.

A second complication is that, regardless of unanticipated changes in usage, O&S costs in each year change over the life cycle of a program. For example, requirements for depot maintenance are typically determined by age or accumulated usage. These costs may be relatively low for several years after a system is initially fielded, but typically rise over time as the fleet ages and accumulates usage. Also, yearly modification costs for individual programs often fluctuate, so actual costs early in the life cycle may not be representative of lifetime costs. Thus, one would expect that actual unit O&S costs for recently fielded systems would be unlikely to capture all the O&S costs expected over the life cycle, which should be captured in SAR estimates.

While acknowledging that it is inherently impossible to determine whether unit O&S cost estimates in SARs accurately predict actual average lifetime unit O&S costs for systems still in service, we compared the most recent available SAR estimates to actual costs to determine whether SAR estimates trend toward a reasonable expectation of actual costs. In making a reasonable expectation of average unit costs over a lifetime, we adjusted predicted usage and costs reported in SARs for actual usage. Two elements of estimated O&S cost—Unit Operations and Maintenance—were adjusted proportionally to actual usage. For example, if actual flying hours per aircraft were one-half of the usage estimated in the SAR for a program, we reduced the estimates for Unit Operations and for Maintenance by one-half when comparing (adjusted) estimated costs with actual costs.

We also considered the point in the life cycle reflected in a program's actual costs. For example, for programs early in their life cycles with long intervals between costly scheduled maintenance, such as overhauls, we considered that the reported yearly maintenance costs to date are unlikely to reflect the average yearly maintenance costs of the program over its lifetime. While we did not adjust estimated or actual costs based on the position of a program in its life cycle, we took this into account when assessing the direction of trends toward reasonable expectations of average lifetime unit costs.

The programs included in this comparison were a subset of the 26 programs for which we tracked O&S estimates. The subset of programs included the 16 Navy and Air Force programs for which we could obtain actual costs: C-130J Hercules transport aircraft, DDG-51 *Arleigh Burke*–class guided missile destroyer, E-2D Advanced Hawkeye electronic aircraft, EA-18G, F-22, F-35, H-1 upgrades, HC/MC-130 Recapitalization, KC-130J, LCS, LPD-17 *San Antonio*–class amphibious transport dock, MH-60R, MQ-9 Reaper, RQ-4A/B Global Hawk, SSN-774, SSN-774 *Virginia*–class submarine, and V-22. We included only direct costs (that is, we excluded element 6.0 Indirect Support) because the estimation and capture of indirect costs in SARs and VAMOSC systems is particularly uneven over time and across VAMOSC systems and MDAPs. After adjusting the predicted costs in SARs for differences between estimated and actual usage, we found the average absolute percentage difference between the latest SAR estimate and recent actual direct O&S costs per unit to be 15 percent.[5] For most of the programs, the latest SAR estimates were higher than actual costs reported in VAMOSC or AFTOC.

We examined the three programs with SAR estimates more than twice the average difference with reported costs and found that they had IOC dates of 2009, 2012, and 2014. Two of the programs had estimated (adjusted for usage) maintenance and modification costs considerably higher than actual costs, which, along with their early stage in the life cycle, suggests that these anticipated costs are likely to be incurred in the future. The third program was for a specific aircraft mission design series.[6] Its estimated costs were lower than the actual costs for the mission design but higher than the actual costs for the mission design series, suggesting that the difference between estimated and actual costs is due to the allocation of actual costs by series.

This comparison of O&S estimates in SARs and actual costs reported in VAMOSC systems says nothing about DoD's or CAPE's current ability to estimate O&S costs. Most of the estimates tracked here were initially made many years ago, long before CAPE was established. The comparison is useful primarily in response to criticisms summarized earlier in this appendix that estimates in SARs tend not to improve over time.

Frequency of Updates of O&S Cost Estimates

For the SARs for which we tracked cost estimates over time, we calculated the percentage of SARs each year with an O&S estimate that changed from the previous year, and the average percentage of the change in the estimate from the previous year.

Before December 2008, the DoD requirement to update O&S estimates in SARs was ambiguous. The estimates for many programs were updated sporadically. DoD guidance directed quarterly updating of sustainment metrics, including cost, in December 2008 (OUSD[AT&L], 2008), which slightly preceded passage of WSARA. In the past several years, officials in OUSD(AT&L) have worked to provide guidance to the services on SARs and ensure the reports are updated with current and correct information.

Table A.3 shows a noticeable increase in the percentage of estimates updated after the guidance. From 1998 to 2007, an average of 64 percent of SARs we checked were updated each year. From 2009 through 2016, an average of 87 percent of the SARs we checked were updated each year. Cost analysts with long experience in DoD have

---

[5] We used the absolute difference between estimated and actual costs so that we could consider SAR estimates higher or lower than actual cost equally. After calculating the percentage difference in absolute values for each program, we calculated the average percentage difference for the 16 programs we assessed.

[6] An example of the mission design series designation is F-16C. The "F" denotes the fighter mission, the "16" denotes the design within the fighter mission, and the "C" denotes the series within the design.

Table A.3
Updates of O&S Cost Estimates

Year of SAR	Percentage of Estimates Updated	Average Percentage Change in Estimate from Last SAR
1998	67	4
1999	50	4
2001	100	3
2002	44	4
2003	67	2
2004	77	3
2005	63	1
2006	59	3
2007	47	2
2009	78	21
2010	95	11
2011	91	6
2012	87	4
2013	92	3
2014	81	0
2015	81	2
2016	89	1

SOURCE: RAND analysis of SAR data.

indicated that, prior to this, some program offices would refrain from updating their cost estimates in SARs for fear of attracting unfavorable attention to the program that could lead to potential budget cuts.[7]

Discussion of O&S Cost Estimates in SARs

In this section of the appendix, we revisit and discuss the findings on the five questions posed at the beginning of the chapter.

---

[7] Multiple cost analysts who worked on MDAPs with large increases in unit O&S cost estimates described this problem. We promised anonymity to interviewees and therefore do not identify the programs, which could jeopardize anonymity.

How much of the life-cycle cost of MDAPs consists of O&S costs? The proportions of O&S costs vary widely according to the type of commodity. For the land, sea, and airborne platforms that generate most of the O&S costs reported for MDAPs, 63 percent of the life-cycle cost consists of O&S costs. For the 82 MDAPs in our sample, 59 percent of the life-cycle cost consists of O&S costs.

Do O&S estimates in SARs become closer to actual costs over time? Yes. One should not expect estimates and actual unit O&S costs to match at a particular point in time, especially for recently fielded systems, because the comparison is of estimated average lifetime unit costs in SARs to actual unit O&S costs at a point in the life cycle. We found a 15-percent difference between the most recent estimated unit O&S costs and recent actual costs in the sample of programs we examined. However, the only findings we view as significant are that the estimated costs do change and trend toward a reasonable estimation over time.

How much have estimates of unit O&S costs changed over time when normalized for usage? The average growth in unit O&S cost for the 20 programs we tracked was 64 percent. The unit O&S costs for the antecedents for these programs, when reported, increased 50 percent. The latter finding is important because the antecedents are fielded systems for which cost data were available when the estimates of the programs of interest were first generated. Much of the growth in the cost of programs and their antecedents was due to personnel compensation, maintenance, and fuel costs that rose faster than inflation.

Is there a difference in the degree of change in the estimates of unit O&S costs in SARs according to the complexity of the weapon system? Yes. Programs that embody significantly new capability or technology, which we categorized as new systems, experience more growth in their estimates of unit O&S cost than do MDAPs that are modifications of existing systems.

Has the frequency of change in SAR O&S estimates changed since WSARA? Yes. WSARA coincided with internal DoD direction to update estimates of O&S costs issued in 2009. There was a sharp increase in the frequency of the updates of estimated O&S costs and in the magnitude of the changes in the 2009, 2010, and 2011 SARs, and the reports continue to be updated more frequently than previously.

Since the beginning of increased management attention in DoD to sustainment costs, which largely coincided with the passage of WSARA, DoD has addressed many of the criticisms of SAR O&S cost estimates made in reports from GAO and others. Overall, the SAR estimates are updated more frequently, provide assumptions and explain changes more consistently, and trend toward reasonableness. These improvements appear to be primarily the result of three changes: improvements in service VAMOSC systems that provide more complete information on the actual costs of analogous systems; OSD direction to the services to update and submit O&S estimates of MDAPs and continued management attention to the subject; and recent laws enacted in NDAAs that focus on sustainment costs, especially the FY 2012 NDAA law that requires the military departments to update their estimates of MDAP O&S costs periodically.

Why Did the O&S Unit Cost Estimates of Some Programs Grow So Much?

In this final section of Appendix A, we seek to shed light on why the estimates of unit O&S costs for some MDAPs grew so much. Toward this end, we reviewed the SARs themselves, including estimates by O&S cost element, explanations of changes, and other supporting text; input data or assumptions in CARDs and LCSPs; cost estimates prepared during development that supported the SAR estimates; data from service databases on reliability, maintainability, and actual numbers of personnel assigned to weapons systems; insights from SMEs knowledgeable about the programs and the estimates for them; and other sources. The information in most of these sources is restricted and cannot be cited in this unrestricted report. Therefore, we refer to the restricted information obliquely but in a way that facilitates understanding of the cost estimates.

We examined the estimates for nine MDAPs that experienced growth in unit O&S cost greater than average. The experience of these programs is not typical of MDAPs:

- C-130J
- CH-53K
- F-22
- F-35
- LCS
- P-8A
- Paladin Integrated Management (PIM)
- RQ-4A/B Global Hawk
- V-22.

We use two kinds of figures to illustrate changes in the estimates. To highlight comparisons between the estimates for a program and for its antecedent, we use line charts. For programs for which no antecedent is shown in the SARs, or to highlight growth in specific elements of O&S cost, we use stacked bar charts. For both kinds of figures, we express the change as a percentage relative to the first unit O&S cost estimate for the program assessed. Missing years in a figure indicate SAR data are missing for those years.

C-130J

The C-130J is the most recent variant of the C-130 cargo aircraft, which first flew in the 1950s. The J variant was developed in the early 1990s at the expense of Lockheed Martin and its suppliers:

> The justification for the new C-130J buys, according to requirements, acquisition, and budget documents, is to reduce the cost of ownership of the C-130E and H fleet, with anticipated cost savings associated with the new technology and the reduced crew and maintenance needs of the J aircraft. (GAO, 1998)

The estimate for C-130J unit O&S costs has nearly quadrupled since the 2003 SAR (see Figure A.1). This was the largest increase of the MDAPs we examined and a far greater increase than for any of the other MDAP modification programs. The growth in the estimate is especially puzzling because actual O&S costs for the C-130H and C-130J have been available throughout this period. Until the sudden increase in the estimate in the last few years, the actual costs have been much higher than the estimated costs for the J variant.

The C-130J SARs did not show an antecedent for the O&S estimate until 2011, when it first showed the C-130H as the antecedent. Actual O&S costs were available and reported in AFTOC for the C-130J and were twice as high as the SAR estimate for the program for several years before the estimate was updated beginning in the 2011 SAR.

Figure A.1
Change in C-130J Unit O&S Cost Estimate

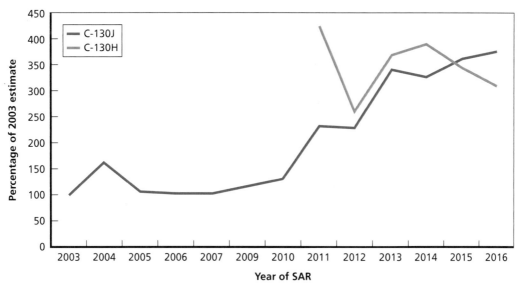

It is possible that higher-than-anticipated usage rates drove part of the increase of the J variant unit O&S cost relative to the H. The C-130J has flown roughly two-thirds more hours per aircraft than the C-130H in FYs 2012 through 2016. However, SARs prior to 2014 do not specify usage rates, so we cannot determine whether usage rates that are higher than originally anticipated affected the increase in unit O&S costs.

CH-53K

The CH-53K heavy lift helicopter provides improvements in lift, range, and other features compared to the CH-53E antecedent it replaces. The SAR estimate for CH-53K unit O&S cost increased by one-half again from the estimate at Milestone B in 2005 (see Figure A.2). Some of the increase was due to the addition of modification costs beginning in the 2010 SAR. But most of the increase is in maintenance costs. There is nothing in the SAR that explains the increase in the unit O&S cost of maintenance.

The estimate for CH-53K unit O&S cost was roughly two-thirds that of its antecedent, the CH-53E, at Milestone B. The CH-53E was reported in the SAR as costing considerably more than actual costs reported in Navy VAMOSC, with the explanation that this represented the projected cost of the CH-53E if it were to continue operation to FY 2053. By the 2016 SAR, the reported cost of the antecedent had dropped to align with the actual costs reported in Navy VAMOSC, and the estimated unit O&S cost of the CH-53K had grown such that it is estimated to cost roughly 40 percent more than its antecedent.

Figure A.2
Change in CH-53K Unit O&S Cost Estimate

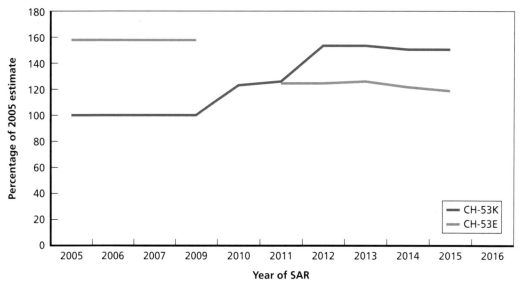

The 2010 SAR for the CH-53K did not report costs for the antecedent, which is why there is a gap in the orange line in Figure A.2.

F-22

The unit O&S cost of the F-22 roughly tripled from the estimate in the 1997 SAR. Most of the cost increase was in the elements of maintenance and unit-level personnel.

The increase in unit-level personnel costs is explained by the estimates of personnel per aircraft provided in the F-22 CARD. The CARD estimated that the F-22 would require roughly 60 percent of the unit-level personnel of the F-15. In FY 2016, the total number of unit-level personnel assigned per aircraft was more than twice the level estimated in the CARD and roughly one-quarter higher than assigned to the F-15.

The optimism persisted as recently as the final SAR for the program in 2010, which reported a current estimate for direct maintenance personnel per aircraft of 9.7. In 2016, AFTOC reported roughly twice as many actual unit-level maintenance personnel per aircraft.

The 2010 SAR similarly reported that the current estimate for mean time between maintenance, 3.0 hours, met the baseline requirement. Because reliability is a driver of maintenance costs, the reported achievement of this key reliability metric makes the sharp increase in estimated maintenance costs contained in the SAR more puzzling.

A service cost position estimate of the program during development showed that estimated unit maintenance costs, including those for the F119 engine, were lower than those for the F-15C. The methodology was based on a program office review and a sufficiency review of the contractor's affordability analysis.

Figure A.3 tracks the estimates of the F-22 and its antecedent F-15C unit O&S cost from 1997 to the final SAR in 2010.

F-35

The SAR estimates the unit O&S cost for the Air Force variant, the F-35A. The estimated unit O&S cost for the F-35A has more than doubled since the beginning of its development program. For several years after the start of development, it was estimated to cost less to fly than its Air Force antecedent, the F-16C. In the 2006 SAR, the estimate for F-35A unit O&S cost increased by two-thirds and, for the first time, was estimated to exceed that of its antecedent. The SAR did not explain the increase.

The unit O&S estimate increased again by roughly one-third in the 2011 SAR. This estimate was produced by CAPE, which borrowed estimators from the services to augment its staff to assist with the estimate. Although the SAR does not explain the increases, most of the increases are in unit-level personnel and maintenance costs. We examined the CARD and cost estimates prepared early in development to understand reasons for the increases in estimated costs since the beginning of development.

Figure A.3
Change in F-22 Unit O&S Unit Cost Estimate

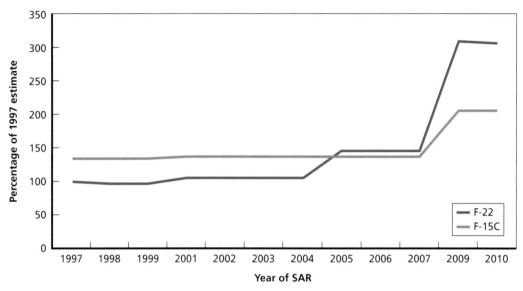

The CARD specified the numbers of unit-level officers and enlisted personnel per squadron, from which we calculated personnel per aircraft and compared the planned figure with actual assigned manning. The number of actual assigned personnel per aircraft in FYs 2014–2016 was roughly two-thirds higher than planned in the CARD. Actual manning efficiency may improve as the fleet matures and as organic maintenance personnel gain proficiency.

The CARD specified threshold and objective values for R&M metrics, including mean flight hours between maintenance events, the key metric for materiel reliability. The FY 2016 DOT&E report shows a threshold value in development of 2.0 mean flight hours between maintenance events for the F-35A as opposed to the observed 1.36 hours during testing (DOT&E, 2016b, p. 90). The estimates for maintenance cost elements were generally based on the cost of legacy aircraft adjusted for the estimated unit recurring flyaway (procurement) cost of the F-35 compared with that of the legacy aircraft and the expected reliability of the F-35. The SAR estimate of unit procurement cost of the F-35 has increased by one-half since development. Given a methodology for F-35 maintenance cost that uses legacy aircraft costs as a baseline and adjusts the baseline cost by the expected reliability and procurement cost of the F-35, the changes experienced to date in F-35 cost and reliability would more than double the maintenance estimate. Using this methodology, any increase in the cost of legacy aircraft maintenance would result in an additional increase in estimated F-35 maintenance costs. Figure A.4 illustrates the change in the F-35A unit O&S cost estimate

Figure A.4
Change in F-35A Unit O&S Cost Estimate

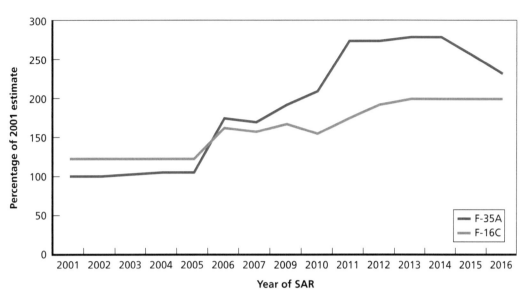

RAND RR2527-A.4

and the cost of its antecedent, the F-16C. The F-16C unit O&S cost increased roughly 60 percent over the period shown.

Littoral Combat Ship

The first LCS was delivered in 2008 and Milestone C approval was given in 2012. The estimate of LCS unit O&S cost increased nearly 60 percent from the 2012 SAR to the 2016 SAR. The increase was driven by the elements of maintenance and unit-level personnel. The reports attribute the increase in unit-level personnel costs to increased crew size. The explanations for the increase in maintenance costs allude to updated requirements for ship availabilities and other maintenance requirements, including shore support. The FY 2014 DOT&E report provides additional insight, describing equipment reliability problems and inadequate training and technical documentation for the ship's crew to allow the crew to isolate equipment failures (DOT&E, 2015, p. 199).

The LCSP for the LCS program illustrates the difficulty for an independent cost estimator to foresee needs for additional funding or to estimate costs based on formal data sources. The LCSP was prepared in 2012 and was approved by the Navy and OSD in 2013, after Milestone C approval. The plan incorporated the R&M and sustainment cost metrics required by OSD guidance and provided threshold and objective values for the metrics. The LCSP also provided a lengthy list of source documents for the metrics provided. The LCSP reported that the Navy had done an independent logistics assessment. Despite the adherence to OSD guidance for establishing and reporting sustainment metrics and producing an approved LCSP, the Navy made fundamental changes in manning and maintenance requirements for the program after Milestone C that led to large increases in unit O&S costs.

The SAR estimates are shown in Figure A.5. The program does not have an antecedent.

P-8

SAR estimates for the P-8 unit O&S cost increased over 80 percent since 2004, with a sizable increase from the 2009 to the 2010 SAR (Figure A.6). Most of the increase was in the elements of maintenance and modifications. Text in the 2010 SAR explained that the increases in the maintenance estimate were due to the addition of reparables costs to the estimate, and an increase in engine depot overhauls from two to three per lifetime. Modification costs increased in 2010 because the estimate included all modifications, not just the safety-of-flight modifications of prior years.

The increase in maintenance costs in the 2016 SAR was explained as due to a variety of factors, including increases for inclusion of capability improvements, updated intermediate-level maintenance manpower, and updated part-level R&M.

Paladin Integrated Management

PIM is an Army program that consists of a self-propelled howitzer and tracked ammunition carrier that provide indirect fire support. The estimate of PIM unit O&S cost increased over 60 percent in two years due to the inclusion of additional elements of cost in the estimates. The increase from 2011 to 2012 was due to the inclusion of train-

Figure A.5
Change in LCS Unit O&S Cost Estimate

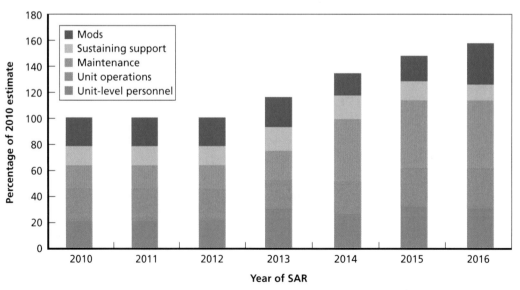

Figure A.6
Change in P-8 Unit O&S Cost Estimate

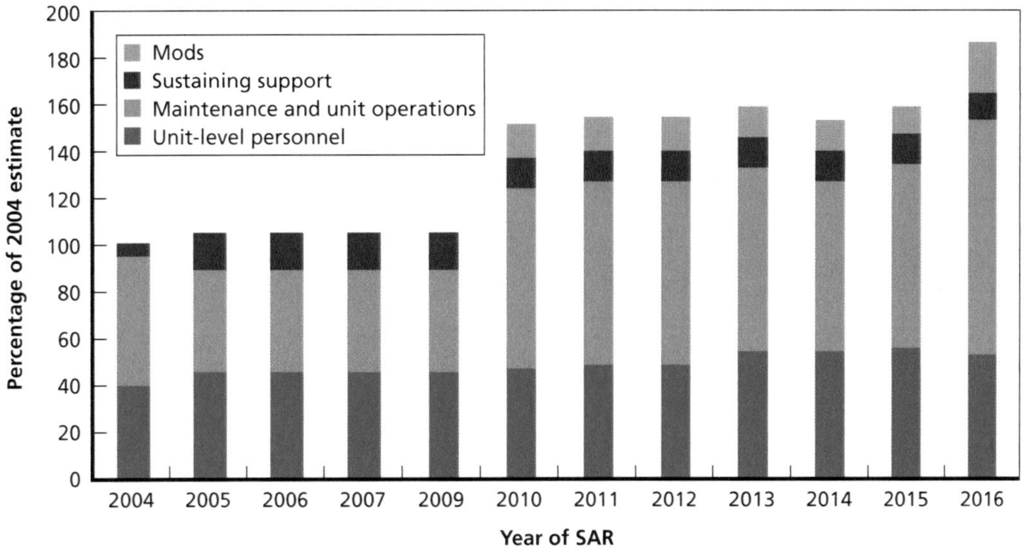

ing ammunition and modifications. The increase from 2012 to 2013 was due primarily to broader inclusion of military personnel costs. It appears the same changes in definition of applicable costs were made in reporting costs of the antecedent system. As shown in Figure A.7, PIM costs slightly exceeded antecedent costs each year.

RQ-4 Global Hawk

Two unusual characteristics of the Global Hawk program provide insight into the growth of its unit O&S cost estimates. First, the program originated as an advanced concept technology demonstration and proved so useful in operation that it was rushed into production without traditional formal development. In their report on support considerations for unmanned aerial vehicles, Drew et al. (2005) discussed how bypassing formal development, when logistics planning is normally done, makes it difficult to determine adequate resources for support.

Second, the Global Hawk program used a spiral development acquisition strategy that added capability in increments, and during the period reported in SARs (2001 to 2014) retired aircraft of the first, less-expensive increment. The original estimate for the program was based on the RQ-4A configuration, but by the end of the period reported in SARs, the estimates reflect a fleet consisting of all RQ-4B aircraft. The RQ-4B variant is larger, has more capability, and has a much higher procurement unit cost. It is reasonable to expect that the unit O&S cost for the B variant would be higher than for the A variant. However, the SAR does not break out O&S costs by variant, so we are unable to adjust or normalize the estimates for this change in the program.

Figure A.7
Change in PIM Unit O&S Cost Estimate

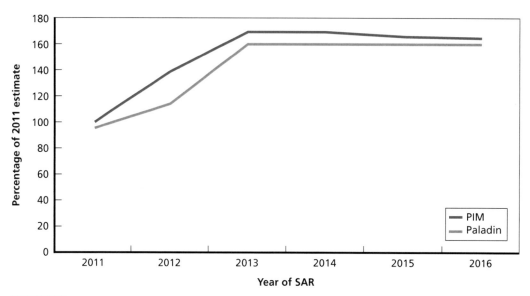

The CARD for Global Hawk specified personnel requirements. The CARD acknowledged Global Hawk's origins as an advanced concept technology demonstration program and associated lack of logistics planning, and highlighted risks to sustainment as a consequence. Actual personnel assigned to the program in FYs 2014 through 2016 averaged 20 percent higher than specified in the CARD.

Figure A.8 shows the increase in Global Hawk unit O&S costs. The narrative in the SARs explains that a source-of-repair assignment process was being performed to determine the long-term depot maintenance strategy, and plans to establish organic depot repair capability were initiated. The source-of-repair analyses were to be completed between 2006 and 2008. The increases in SAR estimates of unit O&S cost followed these analyses. As Drew et al. (2005) discussed, these kinds of logistics planning activities should normally be done earlier in formal development for a traditional acquisition program. The increases in estimated unit O&S costs in the SARs also corresponded with increases in unit procurement costs for more-capable Global Hawk variants.

Actual cost data show that Global Hawk O&S costs per aircraft increased circa 2009 and subsequently dropped. Actual logistics data show an improvement in R&M metrics during this period. These data help explain the changes in the estimates shown in the SARs.

Figure A.8
Change in Global Hawk Unit O&S Cost Estimate

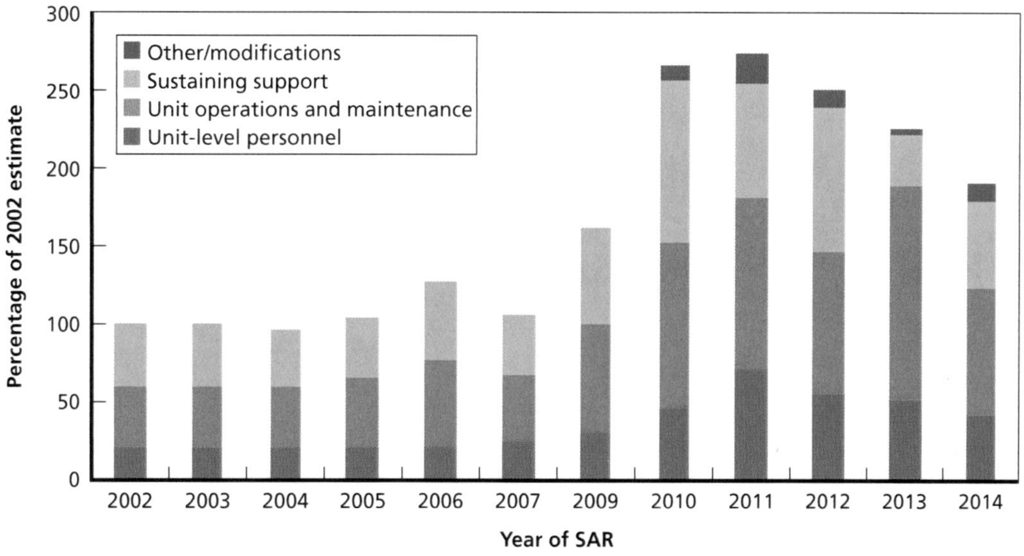

RAND RR2527-A.8

V-22

SARs for the V-22 do not show O&S costs for an antecedent aircraft until 2014, when the CH-46 is shown as the antecedent, and its annual unit O&S cost is presented (Figure A.9). The SAR notes that the antecedent aircraft are not representative of the tilt-rotor V-22. The large increase in unit O&S costs from the 2007 to 2009 SAR is explained as being due to changing the cost-estimating methodology for consumable and reparable parts to base the estimate on actual costs.

The service cost estimate from 2001 provides insight into the earlier estimating methodology for consumable and reparable parts. The V-22 estimates for these elements were based on weight-adjusted costs per flying hour of comparable aircraft, which, notwithstanding the disclaimer in the recent SARs, were legacy helicopters and fixed-wing aircraft. The consumable cost factors were adjusted by expected V-22 reliability. The depot-level reparable cost factors were not adjusted for reliability, and the V-22 depot-level reparable cost per flying hour estimate was almost twice that of the composite average on which it was based.

The Navy VAMOSC system indicates that actual usage of the V-22 has been far below the rates assumed in the SARs. When expressed as a cost per flying hour, actual V-22 costs per flying hour and SAR-estimated costs have nearly converged in recent years.

Figure A.9
Change in V-22 Unit O&S Cost Estimate

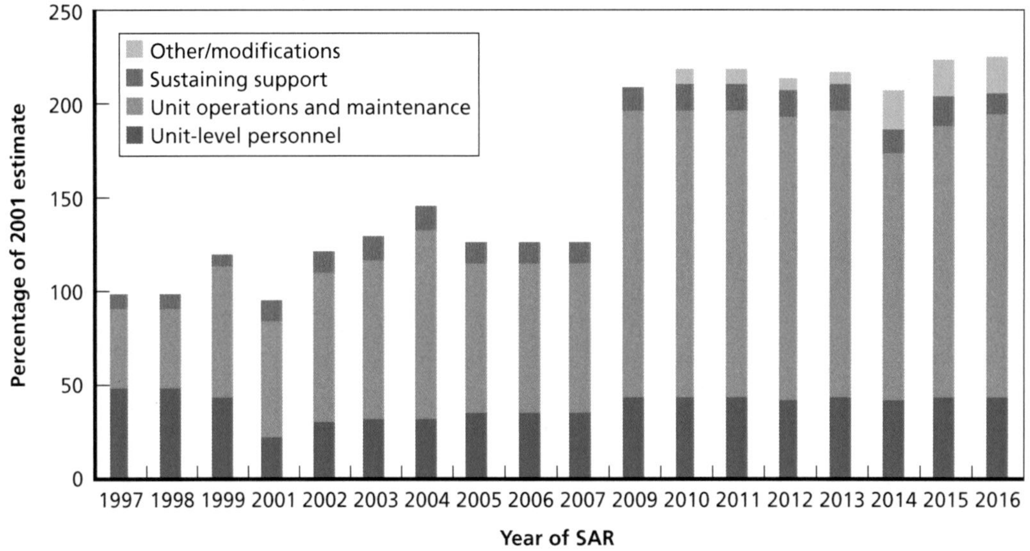

RAND RR2527-A.9

Summary of Reasons for Changes in Estimated Unit O&S Costs and Lessons Learned

This section summarizes the reasons for changes in estimated unit O&S costs. Although a more parsimonious list of explanations would be easier to digest and perhaps more satisfying, we found a variety of reasons for changes in the estimates across the programs we tracked.

One powerful explanation for the increase in estimated unit O&S costs across most, if not all, MDAPs is that the actual O&S costs for antecedent programs increased in constant dollars to an extent cost estimators could not reasonably foresee. Military personnel compensation, fuel, and maintenance costs account for most O&S costs, and these costs have increased faster than the rate of inflation in the general economy since 1997 (the first year for which we have SAR data). Cost analysts draw on historical costs of antecedent programs in developing estimates for new programs and, for O&S costs, project the costs many years into the future. The projections generally assume inflation at officially prescribed rates based on expectations for the general domestic economy. When actual costs for personnel, fuel, and maintenance are higher than forecast for existing systems, this affects new systems too, and estimates of costs for the new systems change accordingly.

Another reason for changes in estimates of unit O&S costs is changes in the definition or scope of the costs included in the estimates. Among the nine MDAPs discussed in earlier in this appendix, the PIM program is a good example of how a decision to include additional elements of cost can lead to above-average changes in cost growth. SARs in recent years contain narrative explanations of such changes, but for most of the years of SAR reporting, little or no explanation is available.

For some programs with higher than average growth, foundational planning documents that inform estimates contained

- estimates of unit-level personnel requirements lower than ultimately assigned
- R&M metrics lower than achieved
- system-level R&M metrics of little use in estimating costs of components or subsystems
- inadequate determination of maintenance requirements.

For programs with higher than average growth, cost-estimating approaches underscore the

- usefulness of antecedent as a sanity check
- risk in relying on assumptions specified in foundational documents without independent validation or sensitivity analysis
- risk of using estimate of procurement cost of new system as a basis for its O&S cost
- value of varying foundational assumptions regarding personnel requirements and R&M to produce a range of estimates
- importance of consistent ground rules for which elements to include in the estimate.

A lesson learned from examining estimates of unit O&S costs in MDAPs is the marked change in the frequency of updates to the estimates, improvement in narrative explanations of changes, and tendency toward realism. These improvements coincide with increased management attention and guidance from OSD, as well as mandates in legislation.

Visibility and Management of Operating and Support Costs Systems

In the body of this report, we emphasized the importance of historical data available in the service VAMOSC systems as a means of validating the accuracy and completeness of prior-year estimates and of developing cost-estimating relationships for future systems. This appendix contrasts the scope of costs included in VAMOSC systems that are directly related to weapon system sustainment with the total sustainment costs available in the VAMOSC systems and with the total cost of sustainment in DoD.

Two appropriations, Military Personnel and O&M, are the primary sources of funds to sustain DoD operations. In FY 2016, DoD spent about $386 billion in these two appropriations ($138 billion for Military Personnel, $248 billion for O&M). The military services receive all the Military Personnel funds. Figure B.1 illustrates the DoD totals for these appropriations for FY 2016.

**Figure B.1
DoD Military Personnel and
Operation and Maintenance
Funding in FY 2016**

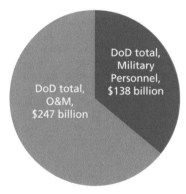

SOURCE: OUSD (Comptroller), 2017.
RAND RR2527-B.1

For O&M, the services receive about 71 percent of the DoD total, while the other 29 percent is allocated among other DoD departments. Figure B.2 shows the allocations of O&M amounts, by military service and all other DoD.

VAMOSC systems over the past 40 years have been collecting O&S cost data specific to weapon systems, but that constitutes only about 24 percent of the total DoD sustainment funding in Military Personnel and O&M appropriations, as shown in Figure B.3.

The VAMOSC systems for the Air Force (AFTOC) and the Navy (Navy VAMOSC) are structured to provide detailed weapon system O&S costs in the CAPE 2014 cost element structure. The Air Force and Navy VAMOSC systems additionally collect virtually all the O&S costs for their respective services in multiple data tables. Although the additional cost data are not structured in the CAPE 2014 format, it is possible to conduct cost and performance analyses for activities not directly associated with weapon system sustainment.

In contrast, the Army Operating and Support Management Information System provides maintenance cost and performance data plus the cost of crew members to operate the systems. The system is not, however, currently structured to capture comprehensive weapon system O&S costs in the CAPE 2014 cost element structure. This system does not include O&S cost data, such as total military manpower, major modifications, training, and other sustaining support. The Army is currently considering methods to integrate data available from other data sources to capture all sustainment costs in the CAPE 2014 cost element structure format.

Figure B.2
DoD Operation and Maintenance Funding in FY 2016

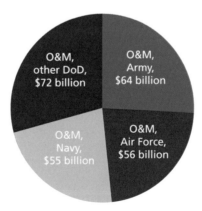

SOURCE: OUSD (Comptroller), 2017.
RAND *RR2527-B.1*

Figure B.3
Operation and Support Funding for Weapon Systems Versus All DoD Operation and Support Funding

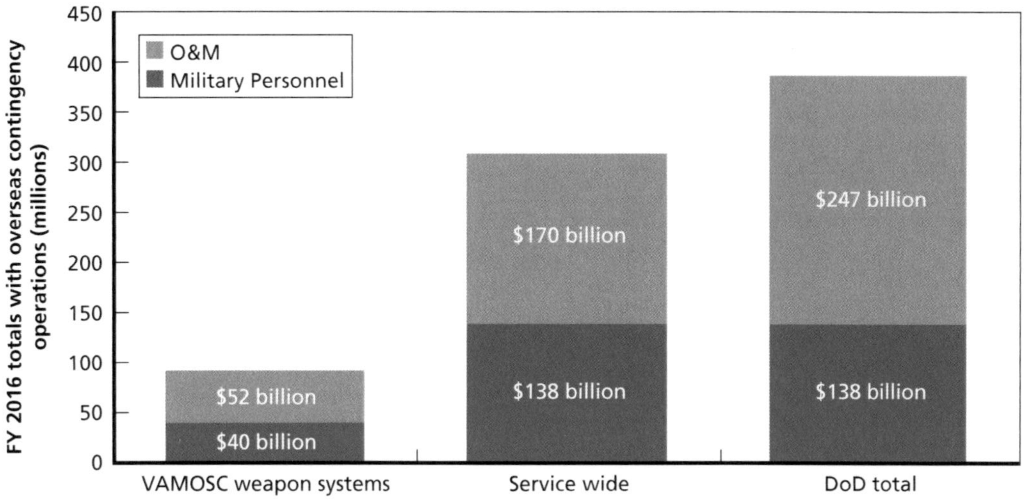

SOURCE: RAND compilation of data from VAMOSC systems and OUSD (Comptroller), 2017.
RAND RR2527-B.3

O&M Funding Not Captured in VAMOSC Systems

In addition to the O&M funds spent by the services, there are 28 DoD directorates that received an additional $77 billion in O&M funding in FY 2016. Figure B.4 illustrates the O&M funding by DoD directorate in FY 2016.

Defense Health Agency (DHA), Washington Headquarters Service (WHS), and Special Operations Command (SOCOM) account for over three-quarters of the total; the other 25 activities account for the remainder.

VAMOSC systems are the primary entities collecting and reporting prior-year costs to operate and support weapon systems. These systems could be broadened to collect and report on additional DoD infrastructure costs that contribute to readiness. As with the reporting of weapon system costs, if paired with appropriate programmatic metrics, the reporting of O&S costs and programmatic outputs for other logistics activities would increase the visibility of their costs and cost trends.

Summary

The three VAMOSC systems receive less than $20 million in funding per year. Because data are either extracted from government financial systems or provided voluntarily from resource sponsors, most of the VAMOSC funding is devoted to database man-

Figure B.4
DoD Operation and Maintenance Funding Not
Captured in VAMOSC Systems

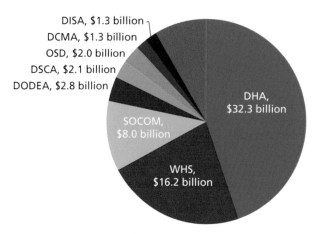

NOTES: DODEA = DoD Education Activity; DSCA = Defense
Security Cooperation Agency; DCMA = Defense Contract
Management Agency; DISA = Defense Information Systems
Agency.
RAND RR2527-B.4

agement and developing data products. Adding additional data sources and generating consistent cost reporting methodologies would require additional resources.

This broad discussion of VAMOSC data is intended to emphasize that direct weapon system O&S costs are only a fraction of DoD O&S cost. Most of DoD's O&S costs are associated with other logistics activities. DoD management of these other logistics support organizations and activities may benefit from the lessons learned by VAMOSC systems and users.

Synopsis of CAPE's Organizational Predecessors, Role, and Functions

CAPE traces its roots to management techniques introduced to OSD by Secretary of Defense Robert McNamara and his staff. While serving as Secretary from 1961 to 1968, McNamara instituted the Planning, Programming, and Budgeting System to centralize these processes. Prior to that centralization, cost information was not collected and analyzed by program, and the military services took the lead in proposing programs. Under McNamara, information was collected by program, and OSD staff assumed the role of evaluating programs and their cost-effectiveness (Hough, 1989).

The Planning, Programming, and Budgeting System aimed to provide a more thorough, analytical, and systematic way of making decisions about force structure, weapon systems, and costs.[1]

To help implement the new management system, Charles J. Hitch, McNamara's comptroller from 1961 to 1965, established an Office of Systems Analysis within the Comptroller Office. The Office of Systems Analysis was headed by Alain C. Enthoven. Systems analysis examines the costs, benefits, and risks of different alternatives for achieving an objective (Fisher, 1970). In 1965, the Office of Systems Analysis was established as the Deputy Assistant Secretary for Systems Analysis, and Enthoven headed the office from September 1965 to January 1969.[2] The office was tasked with cost analysis, which was critical in OSD's examination of alternatives.

With the increased analytical responsibilities given to a new staff of systems analysts in OSD, the military services found that their programs and budgets were coming under more scrutiny. Service leaders tended to resent "what they considered intrusion on their traditional prerogatives" (Trask and Goldberg, 1997, p. 34), and, by the late 1960s, military leaders were publicly criticizing OSD's role in analyzing and making decisions based on the cost-effectiveness of proposed systems (Hough, 1989).

---

[1] For insights into changes in organization and management in DoD and the problems these changes tried to address, see Trask and Goldberg (1997) and Fox et al. (2011)

[2] The new position was established on September 10, 1965, and held initially by Enthoven. Names and dates of tenures of OSD leadership and information about enabling legislation or department guidance can be found in OSD Historical Office, 2016.

In 1971, two years after serving as the head of the Systems Analysis office in OSD, Enthoven and coauthor K. Wayne Smith published the defense classic, *How Much Is Enough: Shaping the Defense Program 1961–1969*.[3] Enthoven and Smith devoted a chapter of the book to answering the question, "Why independent analysts?"—a question that remains relevant today as stakeholders consider roles and responsibilities for acquisition management, including cost estimating, among organizations in DoD. Enthoven and Smith (1971) made three important points in this regard:

- Understanding and exploring fundamental premises (the current term of art is *framing assumptions*) is an important management function in DoD and is "more important than understanding the whole bagful of fancy techniques" (p. 65).
- "Each Service and each important group within a Service is constantly seeking ways to expand its mission and its size and is not immune from using biased assumptions to make its case. Thus, a need remains for an analytic policeman" (p. 107).
- Independent analysts are more free than analysts affiliated with an acquiring organization "to ask hard questions, pose genuine alternatives, and arrive at a recommendation by an objective process" (p. 114).

Melvin Laird became Secretary of Defense in 1969, and, working with his Deputy, David Packard, instituted changes to acquisition processes established under McNamara. First, Laird and Packard returned to the military services the responsibility for identifying needs for weapon systems and for defining, developing, and producing the systems. Hough (1989, pp. 14 and 16) summarizes the transition:

> The shifting emphasis away from systems analysis and the uproar over cost growth permanently changed the role of cost analysis. Where previously cost analysis played a major role in long range planning for analysis of potential-weapon systems, now it would be more important to determine the resource requirements of a proposed weapon system. This change in emphasis from planning to programming to budgeting signaled an urgent requirement for more accurate costing. . . . The role of cost analysis in examining force structures had diminished, but it became an important part in attempting to ensure a more effective procurement system.[4]

A second key change under Laird and Packard was the establishment of the Defense Systems Acquisition Review Council within OSD to advise the Deputy Sec-

---

[3]   The book was republished in 2005 with a new foreword.

[4]   Hough's history of cost analysis in DoD describes the changing and growing role of cost analysis in the DoD acquisition system during this time. The Senate Committee on Armed Services (1985) describes the same shift in management processes and service prerogatives in DoD under McNamara and Laird.

retary on the status and readiness of MDAPs to proceed through each phase of the acquisition life cycle. The policy guidance issued by Packard in 1971 (DoDD 5000.1, 1971) stated that the DoD components are responsible for identifying needs for defense systems and for acquiring them, and that management oversight and reporting requirements should be kept to a minimum. The directive instructed components to request OSD approval to proceed through the acquisition milestones, subject to meeting specified criteria. The same basic process is used today, although reporting requirements have grown since the early 1970s.

In support of the Defense Systems Acquisition Review Council, CAIG was established in 1972 to provide ICEs and to establish uniform DoD cost-estimating standards for use throughout DoD (Senate Committee on Armed Services, 1985). The Defense Systems Acquisition Review Council was replaced by the Defense Acquisition Board in 1987, with the Defense Acquisition Executive as chair (DoDI 5000.01, 1987), but the CAIG retained its roles in providing cost guidance to DoD and cost estimates in support of Defense Acquisition Board decisions. The CAIG was integrated into the CA part of CAPE in 2009, and CA retains these duties.

The position title of Assistant Secretary of Defense (Systems Analysis) was changed to Director of Defense Program Analysis and Evaluation in 1973. The function continued with various changes in the title, position, and emphasis until 2009 (OSD Historical Office, 2016), when WSARA established the Director, CAPE, and transferred the staff of the former office of Program Analysis and Evaluation to CAPE. CAPE's ongoing responsibilities to provide guidance to the services in cost and other analyses reflect the institutionalization of cost and other analytic activities throughout DoD.

Abbreviations

AFTOC	Air Force Total Ownership Cost
AoA	analysis of alternatives
CA	cost assessment
CAIG	Cost Analysis Improvement Group
CAPE	Cost Assessment and Program Evaluation
CARD	cost analysis requirements description
CSDR	cost and software data report
DAMIR	Defense Acquisition Management Information Retrieval
DCMA	Defense Contract Management Agency
DHA	Defense Health Agency
DISA	Defense Information Systems Agency
DoD	U.S. Department of Defense
DoDD	Department of Defense directive
DODEA	DoD Education Activity
DoDI	Department of Defense instruction
DOT&E	Director of Operational Test and Evaluation
DSB	Defense Science Board
DSCA	Defense Security Cooperation Agency
DTC	design-to-cost
EMD	engineering and manufacturing development
FY	fiscal year

GAO	Government Accountability Office (before 2004, General Accounting Office)
ICA	independent cost assessment
ICE	independent cost estimate
IOC	initial operational capability
LCS	littoral combat ship
LCSP	life-cycle sustainment plan
MAIS	major automated information system
MDAP	major defense acquisition program
MIL-STD	military standard
NDAA	National Defense Authorization Act
NRC	National Research Council
O&M	operations and management
O&S	operating and support
OSD	Office of the Secretary of Defense
OUSD(AT&L)	Office of the Under Secretary of Defense for Acquisition, Technology, and Logistics
PIM	Paladin Integrated Management
Pub. L.	Public Law
R&M	reliability and maintainability
SAR	Selected Acquisition Report
SME	subject-matter expert
SOCOM	Special Operations Command
U.S.C.	U.S. Code
VAMOSC	Visibility and Management of Operating and Support Costs
WHS	Washington Headquarters Service
WSARA	Weapon Systems Acquisition Reform Act of 2009

References

Air Force Total Ownership Cost system, undated, not available to the general public. As of October 11, 2017:
https://aftoc.hill.af.mil/

Assistant Secretary of Defense for Research and Engineering, *Technology Readiness Assessment (TRA) Guidance*, April 2011 (rev. May 13, 2011). As of January 8, 2018:
https://www.acq.osd.mil/chieftechnologist/publications/docs/TRA2011.pdf

Blanchard, Benjamin S., *Logistics Engineering and Management*, 6th ed., Upper Saddle River, N.J.: Pearson Prentice Hall, 2004.

Boito, Michael, Thomas Light, Patrick Mills, and Laura H. Baldwin, *Managing U.S. Air Force Aircraft Operating and Support Costs: Insights from Recent RAND Analysis and Opportunities for the Future*, Santa Monica, Calif.: RAND Corporation, RR-1077-AF, 2016. As of January 11, 2018:
https://www.rand.org/pubs/research_reports/RR1077.html

Defense Acquisition Management Information and Retrieval system, undated, not available to the general public.

Defense Science Board, *Report of the Defense Science Board Task Force on Developmental Test and Evaluation,* May 2008. As of October 10, 2017:
http://www.acq.osd.mil/dsb/reports/2000s/ADA482504.pdf

Department of Defense 7000.14R, *Department of Defense Financial Management Regulation (DoD FMR)*, Vol. 6A, Ch. 14, Office of the Under Secretary of Defense (Comptroller), March 2016. As of April 10, 2017:
http://comptroller.defense.gov/Portals/45/documents/fmr/Volume_06a.pdf

Department of Defense Directive 5000.1, *Acquisition of Major Defense Systems*, July 13, 1971. As of May 31, 2017:
http://www.whs.mil/library/mildoc/DODD%205000.1,%2013%20July%201971.pdf

Department of Defense Directive 5000.28, *Design to Cost,* May 23, 1975.

Department of Defense Directive 5105.84, *Director of Cost Assessment and Program Evaluation (DCAPE)*, May 11, 2012. As of May 14, 2018:
http://www.esd.whs.mil/Portals/54/Documents/DD/issuances/dodd/510584p.pdf

Department of Defense Instruction 5000.02, *Operation of the Defense Acquisition System*, January 7, 2015, Incorporating Change 3, effective August 10, 2017. As of May 14, 2018:
http://www.esd.whs.mil/Portals/54/Documents/DD/issuances/dodi/500002_dodi_2015.pdf?ver=2017-08-11-170656-430

Department of Defense Instruction 5000.73, *Cost Analysis Guidance and Procedures*, June 9, 2015. As of June 22, 2017:
http://www.esd.whs.mil/Portals/54/Documents/DD/issuances/dodi/
500073p.pdf?ver=2017-10-02-104030-083

Department of Defense Manual 4140.01, *DoD Supply Chain Materiel Management Procedures: Supply Chain Inventory Reporting and Metrics*, Vol. 10, March 9, 2017. As of August 3, 2017:
http://www.esd.whs.mil/Portals/54/Documents/DD/issuances/414001m/414001m_vol10.pdf

Department of Defense Manual 5000.04, *Cost and Software Data Reporting (CSDR) Manual*, Incorporating Change 1, Effective April 18, 2018. As of May 9, 2018:
http://www.esd.whs.mil/Portals/54/Documents/DD/issuances/dodm/500004p.
pdf?ver=2018-04-18-075226-653

Director, CAPE—*See* Director, Cost Assessment and Program Evaluation.

Director, Cost Assessment and Program Evaluation, *FY 2009 Annual Report on Cost Assessment Activities*, Washington, D.C.: Office of the Secretary of Defense, February 2010a.

———, *Tracking and Assessing Operating and Support Costs for Major Defense Acquisition Programs*, Washington, D.C.: Office of the Secretary of Defense, May 2010b, not available to the general public.

———, *FY 2010 Annual Report on Cost Assessment Activities*, Washington, D.C.: Office of the Secretary of Defense, February 2011.

———, *FY 2011 Annual Report on Cost Assessment Activities*, Washington, D.C.: Office of the Secretary of Defense, February 2012.

———, *FY 2012 Annual Report on Cost Assessment Activities*, Washington, D.C.: Office of the Secretary of Defense, April 2013.

———, *FY 2013 Annual Report on Cost Assessment Activities*, Washington, D.C.: Office of the Secretary of Defense, March 2014.

———, *FY 2014 Annual Report on Cost Assessment Activities*, Washington, D.C.: Office of the Secretary of Defense, February 2015.

———, *FY 2015 Annual Report on Cost Assessment Activities*, Washington, D.C.: Office of the Secretary of Defense, February 2016.

———, *FY 2016 Annual Report on Cost Assessment Activities*, Washington, D.C.: Office of the Secretary of Defense, January 2017.

Director, Operational Test and Evaluation, *FY 2004 Annual Report*, 2004. As of April 16, 2018:
http://www.dote.osd.mil/pub/reports/FY2004/

———, *FY 2007 Annual Report*, December 2007. As of April 16, 2018:
http://www.dote.osd.mil/pub/reports/FY2007/

———, *FY 2013 Annual Report*, January 2014. As of April 16, 2018:
http://www.dote.osd.mil/pub/reports/FY2013/

———, *FY 2014 Annual Report*, January 2015. As of April 16, 2018:
http://www.dote.osd.mil/pub/reports/FY2014/

———, *FY 2015 Annual Report*, January 2016a. As of April 16, 2018:
http://www.dote.osd.mil/pub/reports/FY2015/

———, *FY 2016 Annual Report*, December 2016b. As of April 16, 2018:
http://www.dote.osd.mil/pub/reports/FY2016/

DoD—*See* U.S. Department of Defense.

DoDD—*See* Department of Defense Directive.

DoDI—*See* Department of Defense Instruction.

DOT&E—*See* Director, Operational Test and Evaluation.

DSB—*See* Defense Science Board.

Drew, John G., Russell Shaver, Kristin F. Lynch, Mahyar A. Amouzegar, and Don Snyder, *Unmanned Aerial Vehicle End-to-End Support Considerations*, Santa Monica, Calif.: RAND Corporation, MG-350-AF, 2005. As of October 2, 2017:
https://www.rand.org/pubs/monographs/MG350.html

Enthoven, Alain C., and K. Wayne Smith, *How Much Is Enough? Shaping the Defense Program, 1961–1969*, Santa Monica, Calif.: RAND Corporation, CB-403, 2005. As of September 5, 2017:
https://www.rand.org/pubs/commercial_books/CB403.html

Fisher, Gene H., *Cost Considerations in Systems Analysis*, Santa Monica, Calif.: RAND Corporation, R-490-ASD, 1970. As of March 23, 2017:
http://www.rand.org/pubs/reports/R0490.html

Fox, J. Ronald, with David G. Allen, Thomas C. Lassman, Walton S. Moody, and Philip L. Shiman, *Defense Acquisition Reform, 1960–2009: An Elusive Goal*, Washington, D.C.: U.S. Army Center of Military History, 2011. As of March 23, 2017:
http://history.defense.gov/Portals/70/Documents/acquisition_pub/CMH_Pub_51-3-1.pdf

GAO—*See* General Accounting Office (before 2004) or Government Accountability Office.

General Accounting Office, *Status of the Acquisition of Selected Major Weapon Systems*, Washington, D.C., B163058, February 6, 1970. As of April 12, 2017:
http://archive.gao.gov/otherpdf1/087476.pdf

———, *Intratheater Airlift: Information on the Air Force's C-130 Aircraft*, Washington, D.C., GAO/NSIAD-98-108, April 21, 1998. As of October 3, 2017:
http://www.gao.gov/products/NSIAD-98-108

Government Accountability Office, *DOD Needs Better Information and Guidance to More Effectively Manage and Reduce Operating and Support Costs of Major Weapon Systems*, Washington, D.C., GAO-10-717, July 20, 2010. As of April 26, 2018:
https://www.gao.gov/products/GAO-10-717

———, *Defense Acquisitions: CH-53K Helicopter Program Has Addressed Early Difficulties and Adopted Strategies to Address Future Risks*, Washington, D.C., GAO-11-332, April 4, 2011. As of December 29, 2017:
https://www.gao.gov/products/GAO-11-332

———, *Improvements Needed to Enhance Oversight of Estimated Long- term Costs for Operating and Supporting Major Weapon Systems*, Washington, D.C., GAO-12-340, February 5, 2012. As of May 22, 2017:
http://www.gao.gov/products/GAO-12-340

———, *Acquisition Reform: DOD Should Streamline Its Decision-Making Process for Weapon Systems to Reduce Inefficiencies*, Washington, D.C., GAO-15-192, February 24, 2015. As of April 26, 2018:
https://www.gao.gov/products/GAO-15-192

————, *Weapon System Requirements: Detailed Systems Engineering Prior to Product Development Positions Programs for Success*, Washington, D.C., GAO-17-77, November 17, 2016. As of October 10, 2017:
http://www.gao.gov/products/GAO-17-77

Hough, Paul G., *Birth of a Profession: Four Decades of Military Cost Analysis*, Santa Monica, Calif.: RAND Corporation, P-7539, 1989. As of March 25, 2017:
http://www.rand.org/pubs/papers/P7539.html

House of Representatives Committee on Armed Services, *Report of the Committee on Armed Services House of Representatives on H.R. 2810 Together with Additional Views*, Washington, D.C.: U.S. Government Printing Office, July 6, 2017. As of August 3, 2017:
https://www.congress.gov/115/crpt/hrpt200/CRPT-115hrpt200.pdf

Jones, Gary, Edward White, Erin T. Ryan, and Jonathan D. Ritschel, "Investigation into the Ratio of Operating and Support Costs to Life-Cycle Costs for DoD Weapons Systems," *Defense Acquisition Research Journal*, Vol. 21, No. 1, January 2014, pp. 442–464.

Kneece, R. Royce, Jr., John E. MacCarthy, Jay Mandelbaum, William D. O'Neil, and Gene H. Porter, *Implementing Effective Affordability Constraints for Defense Acquisition Programs*, Alexandria, Va.: Institute for Defense Analyses, P-5123, March 2014. As of July 31, 2017:
https://www.ida.org/-/media/Corporate/Files/Publications/IDA_Documents/SFRD/2014/P-5123.ashx

Lord, Ellen, "Advance Policy Questions for Ellen Lord, Nominee for Under Secretary of Defense for Acquisition, Technology, and Logistics," July 18, 2017. As of July 31, 2017:
https://www.armed-services.senate.gov/imo/media/doc/Lord_APQs_07-18-17.pdf

McCain, John, *Opening Statement by SASC Chairman John McCain at Hearing on DoD Acquisition Reform*, December 7, 2017. As of January 9, 2018:
https://www.mccain.senate.gov/public/index.cfm/2017/12/opening-statement-by-sasc-chairman-john-mccain-at-hearing-on-dod-acquisition-reform

Military Standard 1388-1A, *Logistic Support Analysis*, January 21, 1993.

MIL-STD—*See* Military Standard.

National Research Council, *Industrial Methods for the Effective Development and Testing of Defense Systems*, Washington, D.C.: National Academies Press, 2012.

————, *Reliability Growth: Enhancing Defense System Reliability*, Washington, D.C.: National Academies Press, 2015.

Naval Air Systems Command, 6.0 Logistics and Industrial Operations, "ILA Trend Analysis," March 28, 2017.

Naval Air Systems Command Instruction 5223.2, *Technical/Programmatic Baseline Instruction for NAVAIR Cost Estimates*, August 23, 2010.

NRC—*See* National Research Council.

Office of the Secretary of Defense, Office of Cost Assessment and Program Evaluation, homepage, undated. As of March 23, 2017:
http://www.cape.osd.mil/

Office of the Secretary of Defense, Cost Assessment and Program Evaluation, *Operating and Support Cost Estimating Guide*, March 2014. As of April 13, 2017:
http://www.cape.osd.mil/files/OS_Guide_v9_March_2014.pdf

Office of the Secretary of Defense, Historical Office, *Department of Defense Key Officials: September 1947–October 2016*, October 2016. As of March 23, 2017:
http://history.defense.gov/Portals/70/Documents/key_officials/KEYOFFICIALS-October%202016.pdf?ver=2016-11-29-094819-857

Office of the Under Secretary of Defense for Acquisition, Technology and Logistics, *DoD Weapon System Acquisition Reform Product Support Assessment*, Washington, D.C.: Product Support Assessment Team, November 2009. As of August 2, 2017:
http://www.dtic.mil/dtic/tr/fulltext/u2/a529714.pdf

———, "Implementation of Life Cycle Sustainment Outcome Metrics Data Reporting," December 11, 2008.

———, *Performance of the Defense Acquisition System: 2016 Annual Report*, Washington, D.C., October 24, 2016. As of May 11, 2018:
https://www.acq.osd.mil/fo/docs/Performance-of-Defense-Acquisition-System-2016.pdf

Office of the Under Secretary of Defense (Comptroller), *National Defense Budget Estimates for FY 2017*, C-472D4F8, March 11, 2016. As of March 28, 2018:
http://comptroller.defense.gov/Portals/45/Documents/defbudget/fy2017/FY17_Green_Book.pdf

———, *National Defense Budget Estimates for FY 2018,* D-AA054AD, August 2017. As of March 28, 2018:
http://comptroller.defense.gov/Portals/45/Documents/defbudget/fy2018/FY18_Green_Book.pdf

OUSD(AT&L)—*See* Office of the Under Secretary of Defense for Acquisition, Technology and Logistics.

OUSD (Comptroller)—See Office of the Under Secretary of Defense (Comptroller).

Pub. L.—*See* Public Law.

Public Law 92-156, Armed Forces Appropriation Authorization, 1972, November 17, 1971.

Public Law 94-106, Department of Defense Appropriation Authorization Act, 1976, October 7, 1975.

Public Law 99-145, Department of Defense Authorization Act, 1986, November 8, 1985.

Public Law 104–106, National Defense Authorization Act for Fiscal Year 1996, February 10, 1996.

Public Law 111-23, Weapon Systems Acquisition Reform Act of 2009, May 22, 2009.

Public Law 111–383, Ike Skelton National Defense Authorization Act for Fiscal Year 2011, January 7, 2011.

Public Law 112–81, National Defense Authorization Act for Fiscal Year 2012, December 31, 2011.

Public Law 112–239, National Defense Authorization Act for Fiscal Year 2013, January 2, 2013.

Public Law 113–66, National Defense Authorization Act for Fiscal Year 2014, December 26, 2013.

Public Law 114–92, National Defense Authorization Act for Fiscal Year 2016, November 25, 2015.

Public Law 114–328, National Defense Authorization Act for Fiscal Year 2017, December 23, 2016.

Public Law 115-91, National Defense Authorization Act for Fiscal Year 2018, December 12, 2017.

Recktenwalt, Thomas J., "Visibility and Management of Operating and Support Costs, System II (VAMOSC II)," presentation to the 16th Annual Department of Defense Cost Analysis Symposium, October 4–7, 1981.

Ryan, E., D. Jacques, J. Colombi, C. Schubert, "A Proposed Methodology to Characterize the Accuracy of Life Cycle Cost Estimates for DoD Programs," *Procedia Computer Science*, Vol. 8, 2012, pp. 361–369. As of September 6, 2017:
http://www.sciencedirect.com/science/article/pii/S1877050912000749?via%3Dihub

Ryan, Erin T., David R. Jacques, Jonathan D. Ritschel, and Christine M. Schubert, "Characterizing the Accuracy of DoD Operating and Support Cost Estimates," *Journal of Public Procurement*, Vol. 13, No. 1, Spring 2013, pp. 103–132. As of April 26, 2018:
https://www.emeraldinsight.com/doi/pdfplus/10.1108/JOPP-13-01-2013-B004

Section 809 Panel, *Advisory Panel on Streamlining and Codifying Acquisition Regulations: Section 809 Panel Interim Report*, May 2017.

Senate Committee on Armed Services, *Defense Organization: The Need for Change*, Washington, D.C.: U.S. Government Printing Office, October 16, 1985.

Thornberry, Mac, "Acquisition Reform Legislation," memorandum for House Armed Services Committee members, Washington, D.C., May 18, 2017.

Trask, Roger R., and Alfred Goldberg, *The Department of Defense 1947–1997: Organization and Leaders*, Washington, D.C.: Historical Office, Office of the Secretary of Defense, 1997. As of March 24, 2017:
http://www.dtic.mil/docs/citations/ADA330985

U.S.C.—*See* U.S. Code.

U.S. Code, Title 10, Section 139a, Major Defense Acquisition Program Defined, as of September 10, 2017.

U.S. Code, Title 10, Section 2222, Defense Business Systems: Business Process Reengineering; Enterprise Architecture; Management, as of December 12, 2017.

U.S. Code, Title 10, Section 2321(f), Presumption of Development Exclusively at Private Expense, as of May 1, 2018.

U.S. Code, Title 10, Section 2334, Independent Cost Estimation and Cost Analysis, as of September 10, 2017.

U.S. Code, Title 10, Section 2366a, Major Defense Acquisition Programs: Determination Required Before Milestone A Approval, as of May 1, 2018.

U.S. Code, Title 10, Section 2366b, Major Defense Acquisition Programs: Certification Required Before Milestone B Approval, a as of May 1, 2018.

U.S. Code, Title 10, Section 2430, Major Defense Acquisition Program Defined, as of September 10, 2017.

U.S. Code, Title 10, Section 2432, Selected Acquisition Reports, as of September 10, 2017.

U.S. Code, Title 10, Section 2441, Sustainment Reviews, as of May 1, 2018.

U.S. Code, Title 10, Section 2443, Sustainment Factors in Weapon System Design, as of May 7, 2018

U.S. Code, Title 10, Section 2448a, Program Cost, Fielding, and Performance Goals in Planning Major Defense Acquisition Programs, a as of May 1, 2018.

U.S. Department of Defense, *Selected Acquisition Report (SAR), RCS: DD-A&T(Q&A)823-198, F-35 Joint Strike Fighter Aircraft (F-35)*, December 2001.

U.S. Department of Defense, Office of the Inspector General, *DoD Acquisitions Workforce Reduction Trends and Impacts*, Report No. D-2000-088, February 29, 2000. As of October 11, 2017:
https://media.defense.gov/2000/Feb/29/2001713980/-1/-1/1/00-088.pdf

VAMOSC—*See* Visibility and Management of Operating and Support Cost.

Visibility and Management of Operating and Support Cost database, undated, not available to the general public. As of October 11, 2017:
https://www.vamosc.navy.mil/

WSARA—*See* Public Law 111-23.